SOIL ANALYSIS
Handbook of Reference Methods

SOIL ANALYSIS
Handbook of Reference Methods

Soil and Plant Analysis Council, Inc.

CRC Press
Taylor & Francis Group
Boca Raton London New York

CRC Press is an imprint of the
Taylor & Francis Group, an **informa** business

CRC Press
Taylor & Francis Group
6000 Broken Sound Parkway NW, Suite 300
Boca Raton, FL 33487-2742

First issued in hardback 2017

© 2000 by CRC Press, LLC
CRC Press is an imprint of Taylor & Francis Group, an informa business

No claim to original U.S. Government works

ISBN-13: 9780849302053 (pbk)
ISBN-13: 9781138468672 (hbk)

This book contains information obtained from authentic and highly regarded sources. While all reasonable efforts have been made to publish reliable data and information, neither the author[s] nor the publisher can accept any legal responsibility or liability for any errors or omissions that may be made. The publishers wish to make clear that any views or opinions expressed in this book by individual editors, authors or contributors are personal to them and do not necessarily reflect the views/opinions of the publishers. The information or guidance contained in this book is intended for use by medical, scientific or health-care professionals and is provided strictly as a supplement to the medical or other professional's own judgement, their knowledge of the patient's medical history, relevant manufacturer's instructions and the appropriate best practice guidelines. Because of the rapid advances in medical science, any information or advice on dosages, procedures or diagnoses should be independently verified. The reader is strongly urged to consult the relevant national drug formulary and the drug companies' and device or material manufacturers' printed instructions, and their websites, before administering or utilizing any of the drugs, devices or materials mentioned in this book. This book does not indicate whether a particular treatment is appropriate or suitable for a particular individual. Ultimately it is the sole responsibility of the medical professional to make his or her own professional judgements, so as to advise and treat patients appropriately. The authors and publishers have also attempted to trace the copyright holders of all material reproduced in this publication and apologize to copyright holders if permission to publish in this form has not been obtained. If any copyright material has not been acknowledged please write and let us know so we may rectify in any future reprint.

Except as permitted under U.S. Copyright Law, no part of this book may be reprinted, reproduced, transmitted, or utilized in any form by any electronic, mechanical, or other means, now known or hereafter invented, including photocopying, microfilming, and recording, or in any information storage or retrieval system, without written permission from the publishers.

For permission to photocopy or use material electronically from this work, please access www.copyright.com (http://www.copyright.com/) or contact the Copyright Clearance Center, Inc. (CCC), 222 Rosewood Drive, Danvers, MA 01923, 978-750-8400. CCC is a not-for-profit organization that provides licenses and registration for a variety of users. For organizations that have been granted a photocopy license by the CCC, a separate system of payment has been arranged.

Trademark Notice: Product or corporate names may be trademarks or registered trademarks, and are used only for identification and explanation without intent to infringe.

Visit the Taylor & Francis Web site at
http://www.taylorandfrancis.com

and the CRC Press Web site
http://www.crcpress.com

Library of Congress Cataloging-in-Publication Data

Soil analysis handbook of reference methods / Soil and Plant Analysis Council
    p. cm.
Rev. ed. of: Handbook of reference methods for soil analysis. 1992.
    Includes bibliographical references (p.).
    ISBN 0-8493-0205-6 (alk. paper)
    1. Soils — Analysis — Laboratory manuals. I. Soils and Plant Analysis Council.
S593 .S7415 1999
631.4′1′0287—dc21
                                                                99-048034
                                                                    CIP

Preface

This handbook, *Soil Analysis Handbook of Reference Methods,* is a revision of three previous editions. The first edition, which was published in 1976, was entitled *Handbook on Reference Methods for Soil Testing.* The second and third revisions, both entitled *Handbook of Reference Methods for Soil Analysis,* were published in 1980 and 1992, respectively. The purpose of these *Handbooks* was to "provide a comprehensive and authoritative manual on soil analysis procedures." These *Handbooks* have been extremely well received and have been recognized nationally as well as internationally as standard reference manuals for the commonly used methods of soil analysis.

Since 1976, technological advances in analytical instrumentation and methodology have been substantial. Additionally, the widespread public concern over environmental quality has created a need to expand the coverage of methodology to include methods for elements and constituents not contained in the earlier editions. The preparation and publication of a fourth edition was approved as a recognition of these needs following a recommendation by the Soil and Plant Analysis Council Executive Committee.

For a topic as complex as soil, and for a discipline that contains many diverse soil analysis methods and procedures, it would not be possible for any one individual to prepare such a handbook. Using the third edition of the *Handbook* as a guide, procedures were either deleted, edited, brought up to date, or new methods added. The three regional reference manuals on soil test methodology [see Appendix C (Sims and Wolk, 1991; Gavlak et al., 1994; Brown, 1998)] were referenced to ensure that the methods given in this *Handbook* conform to those that appear in these publications. In order to undertake this task, an Editorial Committee consisting of knowledgeable individuals who were selected based on their subject matter area specialty was formed. These individuals are listed as Contributors on page vi.

The Editorial Committee would also like to express its appreciation to a number of anonymous reviewers who gave their time and talent to ensure that this *Handbook* would maintain its high quality and value to users.

Byron Vaughan
Secretary/Treasurer

Council Officers (1999)

Robert Beck—President
Yash Kalra—Vice President
Byron Vaughan—Secretary/Treasurer
Ann M. Wolf—Past President

Executive Committee (1999)

Todd Cardwell
Stephan J. Donohue
Mark Flock
Karen Gartley
Vince Haby

John Kovar
Richard Large
Manjula Nathan
Cliff Snyder

CAST Representative

Raymond Ward

Contributors

The following individuals assisted in the preparation of the Handbook, either by preparing material on specific methods, or reviewing and editing methods that previously appeared in the 1992 edition.

Edward A. Hanlon, Ph.D.
University of Florida Extension
Immokalee, FL

Gordon V. Johnson, Ph.D.
Department of Agronomy
Oklahoma State University
Stillwater, OK

J. Benton Jones, Jr., Ph.D.
Micro-Macro Publishing
Athens, GA

Yash P. Kalra
Canadian Forestry Service
Northern Forestry Centre
Edmonton, Alberta
Canada

Robert O. Miller, Ph.D.
Colorado State University
Fort Collins, CO

Parviz N. Soltanpour, Ph.D.
Department of Agronomy
Colorado State University
Fort Collins, CO

M. Ray Tucker
North Carolina Department of Agriculture
Agronomic Division
Raleigh, NC

Darryl D. Warnke, Ph.D.
Department of Crop and Soil Sciences
Michigan State University
East Lansing, MI

Maurice Watson, Ph.D.
R.E.A.L., OARDC
Wooster, OH

Table of Contents

Preface v

Contributors vii

Chapter 1
Introduction 1
    1.1   History 1
    1.2   Changing Roles and Needs 3
    1.3   Reference Methods 5
    1.4   North American Proficiency Testing Program (NAPT) 7
    1.5   Reagents, Standards, and Water 10
    1.6   Other Considerations 11
    1.7   Disclaimer 11
    1.8   References 11

Chapter 2
Sampling and Sample Preparation, Measurement, Extraction,
and Storage 17
    2.1   Sampling 17
    2.2   Laboratory Sample Preparation 18
    2.3   Sample Aliquot Determination 20
    2.4   Laboratory Factors 22
    2.5   Long-Term Storage 23
    2.6   References 23

Chapter 3
Soil pH, and Exchangeable Acidity and Aluminum 27
    3.1   pH in Water 28
    3.2   Soil pH in 0.01 M Calcium Chloride 31
    3.3   Soil pH in 1 N Potassium Chloride 33
    3.4   Determination of Exchangeable Acidity Using
           Barium Chloride-TEA Buffer 35
    3.5   Determination of Exchangeable Acidity and Exchangeable
           Aluminum Using 1 N Potassium Chloride 36
    3.6   References 37

Chapter 4
Buffer pH and Lime Requirement 41
    4.1   Adams-Evans Lime Buffer 41

    4.2   SMP Lime Buffer: Original and Double-Buffer Adaptation 45
    4.3   Exchangeable Acidity and Lime Requirement by the
            Mehlich Buffer-pH Method.................................... 49
    4.4   References 54

Chapter 5
Conductance, Soluble Salts, and Sodicity............................... 57
    5.1   Determination of Specific Conductance in
            Supernatant 1:2 Soil:Water Solution 57
    5.2   Determination of Specific Conductance in Supernatant
            1:1 Soil: Water Solution 61
    5.3   Determination of Specific Conductance by Saturated Paste Method ... 64
    5.4   References 66

Chapter 6
Phosphorus 69
    6.1   Bray P1 Extraction.. 70
    6.2   Olsen's Sodium Bicarbonate Extraction 73
    6.3   Mehlich No. 1 Extraction 76
    6.4   Mehlich No. 3 Extraction 80
    6.5   Morgan Extraction.. 83
    6.6   Ammonium Bicarbonate–DTPA Extraction 85
    6.7   References 88

Chapter 7
Major Cations (Potassium, Calcium, Magnesium, and Sodium) 93
    7.1   Neutral Normal Ammonium Acetate Extraction.................. 93
    7.2   Mehlich No. 1 (Double Acid) Extraction 97
    7.3   Mehlich No. 3 Extraction 100
    7.4   Morgan Extraction (Potassium)................................ 104
    7.5   Ammonium Bicarbonate–DTPA Extraction 107
    7.6   Water Extraction 109
    7.7   References 112

Chapter 8
Micronutrients (Boron, Copper, Iron, Manganese, and Zinc)................ 117
    8.1   Hot Water Boron Extraction 118
    8.2   Mehlich No. 1 Zinc Extraction 120
    8.3   Mehlich No. 3 Extraction 123
    8.4   Ammonium Bicarbonate–DTPA Extraction 127
    8.5   DTPA Extraction 130
    8.6   References 133

Chapter 9
Heavy Metals.. 139
   9.1   Ammonium Bicarbonate–DTPA (AB–DTPA) Extraction.......... 139
   9.2   DTPA Extraction 142
   9.3   References 145

Chapter 10
Ammonium- and Nitrate-Nitrogen 149
   10.1   Ammonium Bicarbonate–DTPA Extraction
         (Nitrate Determination)................................. 149
   10.2   2 M Potassium Chloride Extraction (Nitrate and
         Ammonium Determination)................................. 152
   10.3   0.01 M Calcium Sulfate and 0.04 M Ammonium Sulfate
         Extraction (Nitrate Determination) 154
   10.4   References 156

Chapter 11
Sulfate-Sulfur 159
   11.1   Monocalcium Phosphate Extraction 159
   11.2   0.5 M Ammonium Acetate–0.25 M Acetic Acid Extraction 161
   11.3   References 163

Chapter 12
Chloride.. 165
   12.1   0.01 M Calcium Nitrate Extraction 166
   12.2   0.5 M Potassium Sulfate Extraction 168
   12.3   Saturated Calcium Hydroxide Extraction 170
   12.4   References 172

Chapter 13
Organic and Humic Matter 175
   13.1   Wet Digestion 175
   13.2   Loss-On-Ignition (LOI).................................. 178
   13.3   Humic Matter by 0.2 N Sodium Hydroxide Extraction.......... 180
   13.4   References 182

Chapter 14
Organic Soils and Soilless Growth Media Analysis 185
   14.1   Determination of pH, Soluble Salts, Nitrate-Nitrogen, Chloride,
         Phosphorus, Potassium, Calcium, Magnesium, Sodium, Boron,
         Copper, Iron, Manganese, and Zinc in Greenhouse Growth
         Media (Soilless Mixes) by Water Saturation Extraction 185
   14.2   References 190

Chapter 15
Quality Assurance Plans for Agricultural Testing Laboratories 193
- 15.1 Introduction 193
- 15.2 Quality Assurance in the Laboratory 193
  - 15.2.1 Knowing How a Laboratory Operates 194
- 15.3 Statistical Control 196
  - 15.3.1 Statistical Measurements 196
  - 15.3.2 Replicates of Unknowns 198
  - 15.3.3 Spiked Unknowns 198
- 15.4 Control Charts 198
  - 15.4.1 Making Decisions with the Charts. 200
- 15.5 Blind Studies 201
- 15.6 Systems Audits. 202
- 15.7 Documentation. 202
  - 15.7.1 Custody of Samples. 202
  - 15.7.2 Standard Operating Procedures (SOP) and Good Laboratory Procedures (GLP) 203
  - 15.7.3 Methodology References. 203
  - 15.7.4 Instrument Handling and Maintenance Procedures 203
  - 15.7.5 QA Expectations and Measured Performance 204
- 15.8 Summary 205
- 15.9 References 205

Chapter 16
Methods of Instrumental Analysis. 207
- 16.1 Introduction 207
- 16.2 UV-VIS Spectrophotometry 209
  - 16.2.1 Automated UV-VIS Spectrophotometers 211
- 16.3 Flame Emission Spectrophotometry 212
- 16.4 Atomic Absoption Spectrophotometry 213
- 16.5 Inductively-Coupled Plasma Emission Spectrometry 214
  - 16.5.1 Spectrometer Designs 215
    - 16.5.1.1 Sequential Spectrometer. 215
    - 16.5.1.2 Simultaneous Spectrometer 215
    - 16.5.1.3 Photodiode Array Spectrometer 216
  - 16.5.2 Operating Characteristics 216
    - 16.5.2.1 Advantageous Characteristics. 216
    - 16.5.2.2 Disadvantages. 216
    - 16.5.2.3 Standard Preparation 217
    - 16.5.2.4 Calibration Techniques. 217
    - 16.5.2.5 Common Operating Problems 218
    - 16.5.2.6 Important General Points 219
- 16.6 Ion Chromatography (IC) 219

16.7	Specific-Ion Electrodes	220
16.8	References	221

Appendices 225
   A. Reagents, Standards, pH Buffers, Acids, Indicators, Standard Acids, Bases,
      and Buffers Cited in the Handbook 225
   B. Standards and Standard Preparation 231
   C. Conversion Factors 237
   D. Reference Texts 241

Index 245

1 Introduction

This *Soil Analysis Handbook of Reference Methods* is the fourth revision. The first edition was published in 1976 and titled *Handbook on Reference Methods for Soil Testing*. The second and third revisions titled *Handbook of Reference Methods for Soil Analysis* were published in 1980 and 1992 respectively. As in the past, this handbook describes procedures for conducting those soil analysis procedures most frequently used in the U.S., although many of these procedures are also widely used in other parts of the world.

This edition of the handbook retains its initial purpose: *to offer a standard laboratory technique manual for the more commonly used soil analysis (testing) procedures.*

The organization of the text as in the previous editions has been changed as soil analysis procedures are now divided into chapters by element(s) assayed rather than previously organized by method name and/or extraction procedure. Chapters have been added on sampling and sample preparation, determinations for the heavy metals, nitrate and ammonium, chloride, and sulfate, and methods of instrumental analysis.

As in the previous editions, considerable efforts have been made to ensure that sufficient descriptive details are given so that the user will not have to refer to other sources for assistance in conducting the analysis procedure. Pertinent references are given, as appropriate, so that the user can refer back to the source for the origin of the method and/or other historical descriptive material.

In the published literature, a number of terms are used to describe the analysis of soil, the most commonly used term being "soil testing," but in this handbook, the term "soil analysis" is used unless soil testing better describes the topic from a historical point of view.

1.1 HISTORY

In 1942, Piper (1942) published *Soil and Plant Analysis,* a book that is still recognized as a major work on techniques. Piper wrote, "The choice of the analytical method for any given determination depends upon several factors, particularly the purpose for which the analysis is required." Piper continues, "The chemist, called upon to undertake the examination of soils or plant materials without previous experience in this branch of analytical chemistry, is confronted by a vast array of methods in the literature, frequently with little indication of their applicability or relative value." What Piper said 47 years ago still prevails today in the selection of an analytical method for assaying soils.

For more than 60 years, soil testing has been gaining wider use in the U.S. as a basis for determining lime and fertilizer requirements for meeting plant and crop needs (Troug, 1930; Spurway, 1933; Peck et al., 1977, 1990; Voss, 1998). Many of the analysis procedures being used today by public and commercial soil testing laboratories originated in the late 1940s and early 1950s. In a 1953 report, the Soil Test Work Group of the National Soil and Fertilizer Research Committee (Nelson et al., 1953) published a survey of the 50 public soil testing laboratories, finding that there were in use 28 different extractants for determining phosphorus (P) and 19 different extractants for determining potassium (K), many of the methods and techniques having been developed by each laboratory. At that time, the Morgan soil testing method (Morgan, 1932, 1941, 1950; Lunt et al., 1950) was the most widely used soil extractant, particularly in those laboratories located in the northeastern United States and in several western states. The Morgan method has since been modified by Wolf (1982).

In 1973, Jones (1973) examined the soil test methodology in current use and found that considerable standardization in methodology had taken place since the 1953 survey (Nelson et al., 1953). Three extraction methods—Bray P1 (Bray and Kurtz, 1945), Double Acid, currently known as Mehlich No. 1 (Mehlich, 1953), and Olsen (Olsen et al., 1954)—were used to determine P. Similarly, three extraction methods—neutral normal ammonium acetate [(NH$_4$C$_2$H$_3$O$_2$) Schollenberger and Simon, 1945], double acid, now Mehlich No. 1 (Mehlich 1953), and Morgan—were being used to determine extractable K. Although only three extractants existed for determining soil P and K, laboratory techniques varied considerably.

In 1998, another summary of soil analysis methods conducted by Jones found that the Morgan extraction procedure was no longer in wide use as in the past and that the Mehlich No. 3 (Mehlich, 1984), applicable to a wide range of soils, and ammonium bicarbonate (AB)-DTPA (Soltanpour and Schwab, 1977; Soltanpour, 1991) for alkaline soils, were the extractants now in wide use.

In the U.S., four soil test work groups have published summaries of soil analysis procedures in current use in the northeast (Wolf and Sims, 1991), southeast (Donohue, 1992), north-central (Brown, 1998) and western (Gavlak et al., 1994) regions of the U.S. In addition, soil analysis procedures for forest soils have been described by Kalra and Maynard (1991) for Canada, procedures that are widely applicable.

There are five recently published books which describe soil analysis procedures, edited by Westerman (1990) *Soil and Plant Analysis,* Carter (1993) *Soil Sampling and Methods of Analysis,* Sparks (1996) *Methods of Soil Analysis, Part 3 Chemical Methods,* and Peverill et al. (1999) *Soil Analysis: An Interpretative Manual,* and authored by Radojevic and Bashkin (1999) *Practical Environmental Analysis.* For the micronutrients [boron (B), chlorine (Cl), copper (Cu), iron (Fe), manganese (Mn), and zinc (Zn)], Sims and Johnson (1991) have described the analysis procedures for all these elements.

One of the most significant developments that has occurred recently has been the instituting of quality assurance and quality control (QA/QC) programs in soil testing laboratories and the use of soil sample exchange programs, statewide, regionally, and nationally, in the U.S. (Miller et al., 1996) as well as international programs (Houba et al., 1996a). In this handbook, Chapter 15 deals with quality assurance and quality control procedures applicable to the soil analysis laboratory. Hanlon (1996), Klesta and Bartz (1996), and Houba et al. (1996b) have described requirements for a quality laboratory program, while Hortensius and Welling (1996) cover the issues related to the international standardization of soil quality measurements (Quevauviller, 1996). The issue of laboratory certification has been a topic of concern for a number of years, and Blank (1996) discusses this issue as related to laboratory routines, while Jones and White (1996) have described an accreditation program applicable to soil analysis laboratories which was developed by the Soil and Plant Analysis Council (SPAC).

Proficiency testing programs which involve the exchange of carefully prepared soils of specific characteristics have become a significant aspect in a laboratory's QA/QC program. Wolf and Miller (1998) have described the origins of the North American Proficiency Testing Program (NAPT) which is discussed in some detail in Section 1.4 of this chapter. Kalra (1998) has described the Canadian Forest Service's participation in a 25-year check sample program. Houba et al. (1996a) have described the international programme for analytical laboratories (WEPAL) which has its base in Wageningen, The Netherlands (van Dijk et al., 1996).

1.2 CHANGING ROLES AND NEEDS

The farmer/grower whose soil is being properly analyzed and who follows soil analysis-based fertility recommendations can be assured that all lime and fertilizer materials are being correctly and economically applied (Peck and Soltanpour, 1990; McLaughlin et al., 1999) when factors that can affect an interpretation are also considered (Conyers, 1999). Soil testing is big business for public, commercial, and fertilizer company laboratories in the U.S. The most recent survey conducted by the Soil and Plant Analysis Council found that about 5 million soil samples are analyzed annually in the U.S. (Plank, 1998), although the survey numbers are probably an underestimation. The numbers of soils being analyzed during the period from 1958 to 1996 by public and commercial laboratories is shown in Figure 1.1.

As larger numbers of cropped fields are tested, an increasing number of farmers/growers are more fully relying on soil analysis (test) results. The increased interest in soil analysis (testing) is due in part to the increasing cost of fertilizer materials and the desire of farmers/growers to be environmentally correct in their use of agricultural chemicals (Häni, 1996). A more significant role for periodic and systematic soil analysis would be the use of assay results as a monitoring tool for specifying lime and fertilizer treatments needed to establish and then maintain a prescribed soil fertility status, as well as serving as a means of establishing lime and fertilizer needs on regional or national bases, as was recently done by the Potash and Phosphate Institute (Anon., 1998).

If a soil analysis (test) is to be an effective means for evaluating the fertility status of a soil, the standardization of methodology is absolutely essential. No single soil analysis (test) procedure is appropriate for all soils. Specific test procedures have

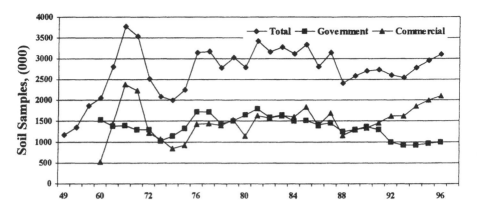

FIGURE 1.1 Number of soil samples assayed by government and commercial laboratories in the U.S. from 1949 to 1996 (Plank, 1998).

been developed for certain soils as one method cannot be used indiscriminately over a wide range of soil-crop conditions. At present, a number of procedures are being used to determine soil pH, lime requirements, and levels of extractable nutrient elements. However, there is some interest in developing more universal test procedures that are adaptable to a wide range of soil-crop conditions (Jones, 1990; van Raij, 1994, 1998; Jones, 1998).

The variance in methodology introduces differences in test results which lead to confusion among farmers/growers and their advisors. Failing to realize that no "single" soil test is applicable to all soils, some may unknowingly send soil samples to a laboratory which does not employ test procedures applicable to their particular soil type. Also, some laboratories fail to realize that their particular soil analysis procedures may not be applicable to all the soils they receive. The result can then lead to a misinterpretation and inappropriate recommendation which may lead to disappointing yields and the possibility of creating a soil fertility imbalance.

Although the number of extractants has significantly narrowed in the last 20 years, techniques vary markedly. In one study, Jones (1973) noted that among those who used the Bray P1 extraction procedure for determining P, soil-to-solution ratios varied from 1:6.7 to 1:10 and shaking times from 40 seconds to five minutes. Bray's (Bray and Kurtz, 1945) original P1 procedure called for a 1:7 soil-to-solution weight ratio with a 40-second shaking time.

Similar differences exist for other test methods. In the extraction of K by neutral normal ammonium acetate, soil-to-solution ratios have varied from 1:3 to 1:10 with various shaking times. Soil pH can be determined in either 1:1 or 1:2 soil-water mixes, in saturated pastes, or in dilute salt solutions (Thomas, 1996). Some laboratories weigh samples, while others scoop to a known volume or an estimated weight. Therefore, the soil to extractant ratio and time of extraction can alter the test result, and thereby affect the interpretation. For those who rely on more than one laboratory to provide soil analysis results, these differences in methodology create an intolerable situation when it comes to evaluating an analysis result.

The various ways of expressing the test result cause confusion, particularly for those who work across state lines or who submit samples to more than one laboratory. Both volume and weight measurements are commonly used for denoting the level of an extracted element. As all too few realize, an error is introduced into the soil analysis result when the volume measurement is used for weighed laboratory samples, or when volume-measured sample results are expressed as parts per million (ppm). To the unsuspecting, differing results for the same soil test procedure may be caused by an erroneous mathematical calculation. Mehlich (1972, 1973) addresses these problems in two practical discussions on volume and weight concepts for expressing a soil analysis result.

Soil test results are also commonly expressed in terms of a sufficiency range, the usual designation denoting a test result as "low," "medium," or "high." Such a coded system is discussed by Cope (1972). A detailed discussion on soil test interpretation may be found in the *ASA Special Publication Number 29* (Peck et al., 1977) and *SSSA Special Publication Number 21* (Brown, 1987), and the book *Soil Testing and Plant Analysis* (revised edition) edited by Westerman (1990).

The actual method of soil analysis is relatively unimportant, compared to the interpretation of the obtained analysis result which defines sufficiency, and particularly for the lime and fertilizer recommendation that is generated by the soil analysis (Cope, 1977; Peck et al., 1977; Brown, 1987; Voss, 1998; McLaughlin et al., 1999). If the laboratory analysis result is well-correlated with crop response or yield, the method of obtaining the analysis result is of minor concern. However, in too many instances laboratory modifications of an already calibrated soil analysis are made without recalibration. For example, many laboratories use the Bray P1 test interpretation data but have modified the original laboratory procedure. In this instance, the modification may alter the analysis result sufficiently to invalidate the interpretation values as originally proposed. The result is a faulty interpretation and an improper P fertilizer recommendation. Therefore, caution is needed when using a soil analysis procedure to ensure that the laboratory methodology is directly related to the associated interpretation. For the Bray P2 test method (Bray and Kurtz, 1945), a stronger acid extraction, was used when rock phosphate was a commonly used P fertilizer source. This P2 test procedure is still used by some laboratories although rock phosphate is no longer in common use (Aquino and Hanson, 1984).

1.3 REFERENCE METHODS

The need to standardize soil analysis procedures and methods is more apparent today, although there is little unanimity on the subject. Several groups are evaluating various analysis methods and setting parameters for each laboratory procedure. In the U.S., much work is being done by regional research committees on soil analysis.

A number of scientific and industrial societies have been engaged in developing and publishing reference methods of analysis. The Association of Official Analytical Chemists (AOAC), which was organized in 1884, is the oldest of these societies in the U.S. The 8th edition of the *Methods of Analysis of AOAC* (Horwitz, 1955) was the last one that included methods of soil analysis. The American Society of Testing Materials (ASTM), the American Public Health Association (APHA, 1989), and, more recently, the Intersociety Committee (Houba et al., 1996c) have been engaged in researching and publishing reference methods of analysis for a wide variety of substances, including soils. The American Society of Agronomy (ASA) and the Soil Science Society of America (SSSA) have published a number of books on methods of soil analysis and interpretation. In 1990, SSSA and AOAC established a joint committee "to conduct validation studies for methods of soil analysis" (Kalra, 1996). The first validation was for soil pH (Kalra, 1995). The Soil and Plant Analysis Council (initially named the Soil Testing and Plant Analysis Council) was organized in 1970 to research and publish reference methods for soil analysis (testing).

The potential and changing role of soil analysis demands reference methods. The growing interest in the environment, the concern by some about overdosing soils with fertilizer, and the need for care in using fertilizer materials in short supply demand more uniformly applied analysis methods. Standardization of methodology is indeed necessary if soil analysis is to be used as a valid monitoring tool. In the near future, regulating agencies may dictate the methods of soil analysis, as various governmental agencies have required AOAC methods for analysis of fertilizers, lime, and other substances (Cunniff, 1999). If soil testing is to become a scientific endeavor, standardization of procedures is essential.

Much of the research published today on crop production contains considerable amounts of soil analysis data which may be of doubtful value. Frequently, these publications do not include references to a particular procedure or provide a detailed description of the method under examination. An article may merely refer to a particular soil test by name without listing the steps used in analysis procedure. To address these discrepancies, this handbook provides a complete description of many different soil analysis procedures, thereby increasing the value of published information found in the methods described in this handbook.

The more commonly used extractants for the major and micronutrients and their application are as follows:

Name	Extractant Composition and Application
Bray P1	0.03 N ammonium fluoride (NH_4F) in 0.025 N hydrochloric acid (HCl); determination of P in acid mineral soils only (Bray and Kurtz, 1945)
Bray P2	0.03 N ammonium fluoride (NH_4F) in 0.1 N hydrochloric acid (HCl); determination of P in acid mineral soils containing sizeable amounts of tricalcium phosphate (Bray and Kurtz, 1945)
Olsen P	0.5 M sodium bicarbonate ($NaHCO_3$); determination of P in alkaline mineral soils only (Olsen et al., 1954)
Neutral Normal Ammonium Acetate	1.0 N ammonium acetate ($NH_4C_2H_3O_2$), pH 7.0; determination of Ca, Mg, K, and Na in all mineral soils (Schollenberger and Simon, 1945)
Morgan	~0.75 N sodium actetate ($NaC_2H_3O_2$), pH 4.8; determination of Ca, Mg, K, and P in most mineral soils (Morgan, 1932, 1941, 1950; Lunt et al., 1950)
Mehlich No. 1	0.05 N hydrochloric acid (HCl) and 0.025 N sulfuric acid (H_2SO_4); determination of Ca, Mg, P, K, Na, and Zn in acid mineral soils low in CEC and organic matter content (sandy acid soils) (Mehlich, 1953)

Introduction

Mehlich No. 3	0.2 N acetic acid ($HC_2H_3O_2$), 0.25 N ammonium nitrate (NH_4NO_3), 0.015 N ammonium fluoride (NH_4F), 0.013N nitric acid (HNO_3), and 0.001 M EDTA; determination of Ca, Mg, P, K, Na, B, and Zn in all acid mineral soils (Mehlich, 1984)
Egnér	0.1 M ammonium lactate and 0.4 M acetic acid ($HC_2H_3O_2$), pH = 3.75; determination of P and major cations (Egnér et al., 1960)
0.01 M Calcium Chloride	0.01 M calcium chloride ($CaCl_2$); determination of P, K, Mg, NO_3, and NH_4, SO_4, Cl, B, Cu, Mn, Fe, Zn, and heavy metals (Houba et al., 1990; Novozamsky et al., 1993)
Ammonium Bicarbonate-DTPA	1 M ammonium bicarbonate (NH_4HCO_3) and 0.005 M DTPA; determination of P, K, Cu, Fe, Mn, Zn, and nitrate in alkaline soils (Soltanpour and Schwab, 1977)
Hot Water Boron	Determination of B in sandy, low organic matter content, acid mineral soils (Berger and Truog, 1944)
DTPA	0.005 M DTPA, 0.1 M TEA, and 0.01 M calcium chloride ($CaCl_2$); determination of Cu, Fe, Mn, and Zn in alkaline soils (Lindsay and Norvell, 1978)

For a soil testing laboratory, only two extraction methods are required, Mehlich No. 3 for acid mineral soils and AB-DTPA for alkaline mineral soils.

In Europe, 0.01 M $CaCl_2$ is becoming the extraction method for all soils.

1.4 NORTH AMERICAN PROFICIENCY TESTING PROGRAM (NAPT)[1]

With increasing attention paid to soil and plant analysis and their role in nutrient management and water quality, there is greater need than ever for soil and plant analysis laboratories to monitor and document the quality of their results. The goal of the North American Proficiency Testing Program (NAPT) is to assist soil and plant tissue testing laboratories in their performance through interlaboratory sample exchanges and a statistical evaluation of the analytical data.

The program operation and sample exchange guidelines have been developed for the agricultural laboratory industry by representatives from groups familiar with and involved in standardizing methods and developing nutrient recommendations for soil and plant analysis methods within the U.S. and Canada. These include: regional soil and plant analysis workgroups; state/provincial departments of agriculture; the Soil and Plant Analysis Council; Soil Science Society of America (SSSA); the Canadian Society of Soil Science; and private and public soil and plant analysis laboratories.

The NAPT program is operated as an activity of the SSSA that includes an oversight committee with representatives from the above-mentioned groups. The program represents a consolidation of seven former soil and plant testing proficiency testing programs.

PROGRAM OBJECTIVES:

- Provide an external quality assurance program for agricultural laboratories
- Develop a framework for long-term improvement of quality assurance of the agricultural laboratory industry
- Identify variability of specific analytical methods

---

[1]Taken directly from the NAPT brochure published in 1999.

SPECIFIC SOIL ANALYSIS:

- Saturated paste percentage (%), pH, EC_e, HCO_3, Ca, Mg, Na, SAR, Cl, SO_4, and B
- Soil pH: (1:1), (1:2) in water and in 0.01 M $CaCl_2$
- Buffer pH-lime requirement (3 methods)
- NO_3-N (5 methods)
- NH_4-N, KCl extractable
- Extractable P (6 methods)
- Extractable K (4 methods) and Al
- Extractable Ca, Mg, and Na (4 methods)
- Extractable SO_4-S, calcium phosphate
- Micronutrients: Zn, Mn, Fe, Cu, B, and Cl (multiple methods)
- Soil organic matter (2 methods)
- Soil total organic carbon and nitrogen
- Inorganic carbon content
- Particle size analysis: sand, silt, and clay
- Cation exchange capacity (CEC)
- KCl extractable Al

SPECIFIC PLANT ANALYSIS:

- NO_3-N, PO_4-P, SO_4-S, and Cl
- Total nitrogen (2 methods)
- Total P, K, S, Ca, Mg, and Na
- Total B, Zn, Mn, Fe, Cu, and Mo (total analysis, 3 methods)

PROGRAM BASICS

The NAPT program is based on the quarterly submission of five soil (600 g) and/or three plant materials (8 g) for chemical analysis using reference methods described in the four Regional Soil Work Group publications (*NEC-67, NCR-13, SERA-6, WCC-103*) and *Methods Manual for Forest Soil and Plant Analysis, Forestry Canada.*

Quarterly program results are statistically compiled and outlier values of individual laboratories identified. Warning limits are reported for each analysis based on the median and median absolute deviation. Results are provided to each participant within 30 days and are kept confidential. An annual report is prepared with an overview of the program quantifying proficiency and reviewing inter- and intralaboratory precision. For laboratories enrolling in the state/province certification programs of Iowa, Illinois, Minnesota, Missouri, Nebraska, Ohio, and Ontario, the NAPT program will provide quarterly laboratory soil results to the respective state program coordinator(s).

FEE STRUCTURE

The 1999 NAPT annual fee is $495 for the soil and plant program and $430 for the soils only program. Exchanges are expected to occur in the months of March, May, August, and October.

Introduction

To enroll in the NAPT Program send payment (payable to SSSA) to:
Soil Science Society of America
Attn.: 1999 NAPT Program
677 South Segoe Road
Madison, WI 53711-1086{clr}

For further information on the NAPT program contact:

Dr. Robert O. Miller
NAPT Program Coordinator
Soil and Crop Sciences Department
Colorado State University
Fort Collins, CO 80523
Phone: 970-493-4382
Fax: 970-416-5820
E-mail: Robm846@aol.com
Rmiller@lamar.colostate.edu

To purchase soils and plant materials:

Janice Kotuby-Amacher
050 Analytical Laboratory
Utah State University
Logan, UT 84332
Phone: 435-797-0008
Fax: 435-797-3376
E-mail: jkotuby@mendel.usu.edu

NAPT PROGRAM BENEFITS

- One of the largest agricultural sample exchange programs in existence (over 160 labs)
- Most comprehensive program available (80 soil and 20 plant standard analysis)
- Collaboration with state/provincial proficiency programs of Iowa, Illinois, Minnesota, Missouri, Nebraska, and Ontario
- Robust statistical estimator techniques used for evaluation of normal and non-normal data sets
- Provides consultation on laboratory methods and quality control to program participants
- Conducts laboratory workshops on methods, techniques, instrumentation, and QC
- Provides for purchase standard reference soil and plant materials utilized in the NAPT program

ACKNOWLEDGMENTS

The North American Proficiency Testing Program is overseen by an oversight committee of the SSSA and acknowledges the contributions of the following groups:

Regional work groups NEC-6T, NCR-13, SERA-6, and WCC-103; Soil and Plant Analysis Council; Soil Science Society of America; Canadian Society of Soil Science; The Minnesota Department of Agriculture; The Missouri Laboratory Testing Program; Nebraska Department of Agriculture; Iowa Department of Agriculture and Land Stewardship; Illinois Soil Testing Association; Missouri Soil Test Certificaton Program; Purdue F5A Program; Ministry of Agriculture Ontario; Michigan State FSA Program; Ohio FSA Program; Wisconsin FSA Program; USDA-NRCS; USEPA; and the commercial laboratory industry.

1.5 REAGENTS, STANDARDS, AND WATER

For all of the analytical procedures described in this handbook, reagents, standards, and water used must be of the highest quality and have characteristics that will not interfere with the analytical procedure.

REAGENTS

A list of all the reagents required to conduct the analytical procedures described in the handbook is found in Appendix A. Reagents should be of *reagent* or *analytical* grade. The storage requirements for many of the reagents are frequently specified to ensure reliable performance. Commercially prepared reagents are sometimes available, particularly extraction reagents, but users are advised to test the quality of these reagents before use.

STANDARDS

The sources, preparations, and testing of standards are described in all of the procedures given in this handbook. For many, the use of commercially prepared standards, whose reliability is quite high, is convenient, saving time both in the preparation and verification testing of user-prepared standards. Therefore, the use of commercially-prepared standards is highly recommended. However, the source and labelling of standards are important considerations, to ensure freedom from analytes in a standard that may be included in a multielement assay and ensure that the characteristics of the matrix (mix of cations and anions), and acid content [nitric (HNO_3) or hydrochloric (HCl), or both] will not affect or interfere with the analytical procedure being used. The preparation and use of standards is discussed in some detail in Appendix B.

WATER

The quality of water used in the preparation of reagents and standards is critical to ensure reliability of the analytical procedure conducted. When the word *water* is used in this handbook, it refers to pure water, water free from any dissolved ions or other substances. Such water may be obtained commercially or by means of distillation (single or double), ion exchange, and/or reverse osmosis (Anon., 1997). The water used should be tested using those analytical procedures by which the presence of low ion concentration can significantly affect the analytical result. One example is determination of P by the molybdenum blue spectrophotometric procedure where low levels of the arsenate (AsO_4^{2-}) and/or silicate (SiO_4^{4-}) anion can generate the same blue color as that of the orthophosphate (PO_4^{3-}) anion.

1.6 OTHER CONSIDERATIONS

The ruggedness of an analytical procedure, that is, the exactness required for each parameter, i.e., tolerance for variance, is becoming of increasing importance and the parameters given with each procedure in this handbook should be strictly followed to ensure reliable performance of the method. Factors, such as the condition of the assayed sample, pH and composition of reagents, time, temperature, physical parameters in terms of shaking speeds, characteristics of storage and extraction vessels, weight and volume measurements of the samples, reagents, and standards, instrument settings, and methods of instrument calibration and operation, are normally specified and should be exactly followed. What might be perceived as an acceptable variance by an analyst can invalidate the analytical result obtained.

In most instances, a dual system of weighed and/or volume measured samples is presented. This rationale is necessary in cases in which the original method carefully specified a weight of sample or volume of known or assumed specific weight. The reader may refer to Mehlich (1972, 1973) for additional information on volume-weight considerations and to Peck (1998) for more details on scoop design and use.

Verification of analytical results by applying the principles of quality assurance and quality control, frequently referred to as QA/QC laboratory procedures, is discussed in considerable detail in Chapter 15. Participation in a proficiency testing program provides one means of ensuring reliable laboratory performance. The North American Proficiency Testing program is described in some detail in Section 1.4 of this introductory chapter.

For those looking for analytical assistance, the recently published *Registry of Soil and Plant Analysis Laboratories in the United States and Canada* (CRC Press, Inc., 2000 Corporate Blvd., N.W., Boca Raton, FL 33431-7737) provides a listing of laboratories, including information on analytical services provided, contact person, etc.

1.7 DISCLAIMER

The naming of products in this handbook does not constitute endorsement. The analysis procedures described in this handbook have been adopted from current publications found in the public domain.

1.8 REFERENCES

Anon. 1997. Water: What's in it and how to get it out. *Today's Chem.* 6(1):16–19.
Anon. 1998. *Soil Test Levels in North America: Summary Update.* PPI/PPIC/FAR Technical Bulletin 1998–3. Norcross, GA: Potash & Phosphate Institute.
APHA. 1989. *Standard Methods for the Examination of Water and Wastewater,* 17th ed. Washington, DC: American Public Health Association.
Aquino, B. F. and R. G. Hanson. 1984. Soil phosphorus supplying capacity evaluated by plant removal and available phosphorus extraction. *Soil Sci. Soc. Amer. J.* 48:1091–1096.

Berger, K. C. and E. Truog. 1944. Boron tests and determination for soils and plants. *Soil Sci.* 57:25–36.

Blank, F. T. 1996. Certification aspects in laboratories for soil and plant analysis. *Commun. Soil Sci. Plant Anal.* 27:349–363.

Bray, R. H. and L. T. Kurtz. 1945. Determination of total, organic, and available forms of phosphorus in soils. *Soil Sci.* 59:39–45.

Brown, J. R. (ed.). 1987. *Soil Testing: Sampling, Correlation, Calibration, and Interpretation.* SSSA Special Publication 21. Madison, WI: Soil Science Society of America.

Brown, J. R. (ed.). 1998. *Recommended Chemical Soil Test Procedures for the North Central Region.* North Central Regional Publication 221 (revised). Missouri Agricultural Experiment Station Bulletin SB 1001. Columbia, MO: University of Missouri.

Carter, R. L. (ed.). 1993. *Soil Sampling and Methods of Analysis.* Boca Raton, FL: CRC Press.

Conyers, M. K. 1999. Factors affecting soil test interpretation, In: *Soil Analysis: An Interpretation Manual,* K. I. Peverill, L. A. Sparrow, and D. J. Reuter (eds.), CSIRO Collingwood, Australia: CSIRO Publishing, pp. 23–34.

Cope, Jr., J. T. 1972. *Fertilizer Recommendations and Computer Program Key Used by the Soil Testing Laboratory.* Auburn Experiment Station Circular 176.

Cunniff, P. A. (ed.). 1989. *Official Methods of Analysis of the Association of Official Analytical Chemists,* 16th ed. Volume 1. Arlington, VA: Association of Official Analytical Chemists.

Donohue, S. T. (ed.). 1988. *Reference Soil Test and Media Diagnostic Procedures for the Southern Region of the United States.* Southern Cooperative Series Bulletin. Blacksburg, VA:VPI.

Egnér, H. H. Riehm, and W. R. Domingo. 1960. Untersuchungen über die chemische Bodenanalyse als Grundlage für die Beurteilung des Nährtoffzustandes der Böden. II. Chemische Extraktionsmethoden zur Phosphor- und Kalium-bestimmung. *Kungl. Lanthr. Hogsk. Ann.* 26:199–215.

Gavlak, R. G., D. A. Horneck, R. O. Miller (eds.). 1994. *Plant, Soil, and Water Reference Methods for the Western Region.* Western Regional Extension Publication WREP 125. Fairbanks, AK: University of Alaska Cooperative Extension Service.

Häni, H. 1996. Soil analysis as a tool to predict effects on the environment. *Commun. Soil Sci. Plant Anal.* 27:289–306.

Hanlon, E. 1996. Laboratory quality: A method of change. *Commun. Soil Sci. Plant Anal.* 27:307–325.

Hortensius, D. and R. Welling. 1996. International standardization of soil quality measurements. *Commun. Soil Sci. Plant Anal.* 27:387–402.

Horwitz, W. (ed.). 1955. *Methods of Analysis of the AOAC.* 8th Edition. Washington, DC: Association of Official Agricultural Chemists.

Houba, V. J. G., I. Novozamsky, Th.M. Lexmond, and J. J. van der Lee. 1990. Applicability of 0.01 M $CaCl_2$ as a single extractant for the assessment of the nutrient status of soils and other diagnostic purposes. *Commun. Soil Sci. Plant Anal.* 21:2281–2290.

Houba, V. J. G., J. Uittenbogaard, and P. Pellen. 1996a. Wageningen evaluating programmes for analytical laboratories (WEPAL): Organization and purpose. *Commun. Soil Sci. Plant Anal.* 27:421–431.

Houba, V. J. G., I. Novozamsky and J. J. van der Lee. 1996b. Quality aspects soil in laboratories for soil and plant analysis. *Commun. Soil Sci. Plant Anal.* 27:327–348.

Houba, V. J. G., Th. M. Lexmond, I. Novozamsky, and J. J. van der Lee. 1996c. State of the art and future developments in soil analysis for bio-availability assessment. *Sci. Total Env.* 178:21–28.

Jones, Jr., J. B. 1973. Soil testing in the United States. *Commun. Soil Sci. Plant Anal.* 4:307–322.

Jones, Jr., J. B. 1990. Universal soil extractants: Their composition and use. *Commun. Soil Sci. Plant Anal.* 21:1091–1101.

Jones, Jr., J. B. 1998. Soil test methods: Past, present, and future use of soil extractants. *Commun. Soil Sci. Plant Anal.* 29:1543–1552.

Jones, Jr., J. B. and W. C. White. 1994. An accreditation program for soil and plant analysis laboratories. *Commun. Soil Sci. Plant Anal.* 25:843–857.

Kalra, Y. P. 1995. Determination of pH of soils by different methods: Collaborative study. *J. Assoc. Off. Anal. Chem.* 78:310–324.

Kalra, Y. P. 1996. Soil pH: First soil analysis methods validated by the AOAC International. *J. For. Res.* 1:61–64.

Kalra, Y. P. 1998. Canadian Forest Service's participation in check sample programs over a 25-year period. *Commun. Soil Sci. Plant Anal.* 28:1667–1684.

Kalra, Y. P. and D. G. Maynard. 1991. *Methods Manual for Forest Soil and Plant Analysis.* Northwest Region Information Report NOR-X-319. Forestry Canada, Canadian Forest Service, Edmonton, Canada.

Klesta, E. J. and J. K. Bartz. 1996. Quality assurance and quality control, In: *Methods of Soil Analysis, Part 3, Chemical Methods,* SSSA Book Series No. 5. D.L. Sparks (ed.). Madison, WI: Soil Science Society of America. 19–48.

Lindsay, W. L. and W. A. Norvall, 1978. Development of a DTPA micronutrient soil test for zinc, iron, manganese, and copper. *Soil Sci. Soc. Amer. J.* 42:421–428.

Lunt, H. A., C. L. W. Swanson, and H. G. M. Jacobson. 1950. *The Morgan Soil Testing System.* Connecticut Agricultural Experiment Station Bulletin 541:1–60.

McLaughlin, D. J. Reuter, and G. E. Rayment. 1999. Soil Testing—Principles and concepts, In: *Soil Analysis: An Interpretation Manual,* K. I. Peverill, L. A. Sparrow, and D. J. Reuter (eds.). Collingwood, Australia: CSIRO Publishing. 1–21.

Mehlich, A. 1953. *Determination of P, Ca, Mg, K, Na, and NH_4.* Raleigh, NC: Mimeo, North Carolina Soil Testing Division.

Mehlich, A. 1972. Uniformity of expressing soil test results: A case for calculating results on a volume basis. *Commun. Soil Sci. Plant Anal.* 3:417–424.

Mehlich, A. 1973. Uniformity of soil test results as influenced by volume weight. *Commun. Soil Sci. Plant Anal.* 4:475–486.

Mehlich, A. 1984. Mehlich 3 soil test extractant: A modification of Mehlich 2 extractant. *Commun. Soil Sci. Plant Anal.* 15(12):1409–1416.

Miller, R. O., J. Kotuby-Amacher, and N.B. Dellavalle. 1996. A proficiency testing program for the agricultural laboratory industry: Results of the 1994 program. *Commun. Soil Sci. Plant Anal.* 27:451–461.

Morgan, M. F. 1932. *Microchemical Soil Tests.* Connecticut Agricultural Experiment Station Bulletin 333.

Morgan, M. F. 1941. *Chemical Diagnosis by the Universal Soil Testing System.* Connecticut Agricultural Experiment Station Bulletin 450.

Morgan, M. F. 1950. *Chemical Diagnosis by the Universal Soil Testing System.* Connecticut Agricultural Experiment Station Bulletin 451.

Nelson, W. L., J. W. Fitts, L. D. Kardos, W. T. McGeorge, R. Q. Parks, and J. Fielding Reed. 1953. *Soil Testing in the United States,* 0–979953. Washington, DC: National Soil and Fertilizer Research Committee, U.S. Government Printing Office.

Novozamsky, I., Th. M. Lexmond, and V. J. G. Houba. 1993. A single extraction procedure of soil for evaluation of uptake of some heavy metals by plants. *Int. J. Environ. Anal. Chem.* 51:47–58.

Olsen, S. R., C. V. Cole, F. S. Watanabe, and L. A. Dean. 1954. *Estimation of Available Phosphorus in Soils by Extraction with Sodium Bicarbonate.* USDA Circular No. 939. Washington, DC: U.S. Government Printing Office.

Peck, T. R. and J. T. Cope, Jr., and D. A. Whitney (eds.). 1977. *Soil Testing: Correlating and Interpreting the Analytical Results.* ASA Special Publication No. 19. Madison, WI: American Society of Agronomy.

Peck, T. R. 1998. Standard soil scoop. In: *Recommended Chemical Soil Test Procedures for the North Central Region,* J. R. Brown (ed.). North Central Regional Publication 221 (revised). Missouri Agricultural Experiment Station SB 1001. Columbia, MO: University of Missouri, 7–9.

Peck, T. R. 1990. Soil testing: Past, present, and future. *Commun. Soil Sci. Plant Anal.* 21:1165–1186.

Peck, T. R. and P. N. Soltanpour. 1990. The principles of soil testing, In: *Soil Testing and Plant Analysis,* 3rd ed. SSSA Book Series No. 3. R. L. Westerman (ed.). Madison, WI: Soil Science Society of America, 1–9.

Peverill, K. I., L. A. Sparrow, and D. J. Reuter (eds.). 1999. *Soil Analysis: An Interpretation Manual.* CSIRO Publishing, Collingwood, Australia.

Piper, C. S. 1942. *Soil and Plant Analysis.* Adelaide, Australia: The University of Adelaide.

Plank, O. (ed.). 1998. *Soil, Plant and Animal Waste Analysis: Status Report for the United States, 1992–1996.* Lincoln, NE: Soil and Plant Analysis Council.

Quevauviller, P. 1996. Certified reference materials for the quality control of total and extractable trace element determinations in soils and sludges. *Commun. Soil Sci. Plant Anal.* 27:403–418.

Radojevic, M. and V. N. Bashkin. 1999. *Practical Environmental Analysis.* Cambridge, U.K.: Royal Society of Chemistry.

Schollenberger, C. J. and R. H. Simon. 1945. Determination of exchange capacity and exchangeable bases in soil—ammonium acetate method. *Soil Sci.* 59:13–24.

Sims, J. T. and G. V. Johnson. 1991. Micronutrient soil tests. In: *Micronutrients in Agriculture,* 2nd ed. SSSA Book Series No. 4. J. J. Mortvedt (ed.). Madison, WI: Soil Science Science of America, 427–476.

Sparks, D. L. (ed.). 1996. *Methods of Soil Analysis, Part 3, Chemical Methods.* SSSA Book Series Number 5. Madison, WI: Soil Science Society of America.

Spurway, C. H. 1933. *Soil testing: A Practical System of Soil Diagnosis.* Michigan Agricultural Experiment Station Bulletin 132.

Soltanpour, P. N. 1991. Determination of nutrient availability and elemental toxicity by AB-DTPA soil test and ICPS. *Adv. Soil Sci.* 16:165–190.

Soltanpour, P. N. and A. P. Schwab. 1977. A new soil test for simultaneous extraction of macro- and micro-nutrients in alkaline soils. *Commun. Soil Sci. Plant Anal.* 8:195–207.

Thomas, G. W. 1996. Soil pH and soil acidity, In: *Methods of Soil Analysis, Part 3, Chemical Methods,* SSSA Book Series Number 5. D. L. Sparks (ed.). Madison, WI: Soil Science Society of America. 475–490.

Truog, E. 1930. The determination of the readily available phosphorus in soils. *J. Amer. Soc. Agron.* 22:874–882.

van Dijk, D., V. J. G. Houba, and J. P. J. van Dalen. 1996. Aspects of quality assurance within the Wageningen evaluating programmes for analytical laboratories. *Commun. Soil Sci. Plant Anal.* 27:433–439.

van Raij, B. 1994. New diagnostic techniques, universal soil extractants. *Commun. Soil Sci. Plant Anal.* 25:799–816.

van Raij, B. 1998. Bioavailable tests: Alternatives to standard soil extractions. *Commun. Soil. Sci. Plant Anal.* 29:1553–1570.

Voss, R. 1998. Fertility recommendations: Past and present. *Commun. Soil Sci. Plant Anal.* 29:1429–1440.

Westerman, R. L. (ed.) 1990. *Soil Testing and Plant Analysis,* 3rd ed. SSSA Book Series No. 3. Madison, WI: Soil Science Society of America.

Wolf, A. M. and J. T. Sims (eds.) 1991. *Recommended Soil Testing Procedures for the Northeastern United States.* Northeast Regional Publication No. 493. Newark, DE: Agricultural Experiment Station, University of Delaware.

Wolf, A. M. and R. O. Miller. 1998. Development of a North American proficiency testing program for soil and plant analysis. *Commun. Soil Sci. Plant Anal.* 28:1685–1690.

Wolf, B. 1982. An improved universal extracting solution and its use for diagnosing soil fertility. *Commun. Soil Sci. Plant Anal.* 13:1005–1033.

2 Sampling and Sample Preparation, Measurement, Extraction, and Storage

The value of a soil analysis result is no better than the quality of the sample assayed, determined by (1) how the example is taken from the field (James and Wells, 1990; Crépin and Johnson, 1993; Peterson and Calvin, 1996; Reid, 1998; Peck and Beck, 1998; Schnug et al., 1998; Wright, 1998; Brown, 1999; Radojevic and Baskin, 1999), (2) what conditions existed in transport to the laboratory, (3) the type of preparation techniques used to prepare the laboratory sample (Bates, 1993; Anon., 1994; Hoskins and Ross, 1995; Gelderman and Mallarino, 1998; Brown, 1999; Radojevic and Baskin, 1999), (4) sample aliquot measurement (Mehlich, 1972, 1973; van Lierop, 1981, 1989; Bates, 1993; Peck, 1998), (5) laboratory factors (Eliason, 1998), and (6) sample storage (Bates, 1993; Houba and Novozamsky, 1998; Brown, 1999). In this chapter, techniques commonly used to obtain the proper sample and those procedures required to maintain sample integrity in transit, during laboratory preparation, and analysis are described.

2.1 SAMPLING

1. Field Sampling

Soils are naturally variable horizontally as well as vertically, requiring careful consideration in terms of sampling technique. Topography and soil type are common factors for determining sampling boundaries for collecting a single soil composite. There are three commonly used sampling strategies:

- Simple Random Sampling
- Stratified Random Sampling — selecting individual soil cores in a random pattern within a designated area
- Systematic or Grid Sampling

There are statistical concepts in soil sampling that will determine which method of sampling best defines the area under test evaluation. Since any detailed discussion is beyond the scope of this handbook, readers are referred to the review articles on this topic by Sabbe and Marx (1987), James and Wells (1990), Crépin and Johnson (1993), Peterson and Calvin (1996), Peck and Beck (1998), Radojevic and Baskin (1999), and Brown (1999) for general sampling considerations and to Schnug et al. (1998), Nowak (1998), Wright (1998), and Anon. (1999) for systematic or grid sampling procedures.

The depth of sampling is determined by one of several factors (Brown, 1999): horizonal characteristics (limiting depth to one soil horizon), depth of soil mixing for land preparation, and rooting depth of crop growing or to be grown.

The number of cores needed to constitute a single sample composite is not a fixed number but will vary depending on site variability and statistical factors. In general, the number of cores recommended to constitute a composite sample ranges from 10 to 20 cores.

The devices used to collect soil sample cores and the type of vessels and/or transport containers can contaminate a sample with a to-be-determined analyte. For example, sampling devices and/or mixing containers made of brass will add copper (Cu) and zinc (Zn), galvanized metal and rubber will add Zn, and moist soil contact with iron implements will add iron (Fe). The glue in some types of paper bags, if it becomes moist from wet soil, can add boron (B). Therefore, care needs to taken to ensure that a collected soil sample is not brought into contact with substances that will affect sample integrity, or be exposed to an atmosphere that will add foreign substances (i.e., dust) to a collected sample. A review of procedures for preventing contamination has been given by James and Wells (1990).

Various core sampling devices are available, both hand- or power-driven. The latter can be mounted on vehicles for ease of movement through the field and taking cores. Some suppliers for these devices are listed at the end of this chapter.

Detailed descriptions of soil sampling procedures, as well as other useful information on soil sample handling, can be found on the Internet.

2. Transport to the Laboratory

If the period of time between field sample collection and arrival at the laboratory will be more than several days, field-moist soil, when placed in an air-tight container, can undergo significant biological changes at room and/or elevated temperatures. Organic matter decomposition can release elements (ions) such as phosphorus (P), sulfate (SO_4), boron (B), and nitrate (NO_3) into the soil solution, while anaerobic conditions can result in organic matter decomposition and loss of nitrogen (N) from the soil. For long-term transport, the collected soil should be kept in a cool environment [5–10°C (40–50°F)] and excess water should be removed by partial drying, keeping the soil just moist.

Freezing a soil sample will maintain soil biological integrity, but it may significantly alter the physio-chemical properties, as freezing has the same effect on soil as high-temperature [>32°C (>90°F)] drying.

2.2 LABORATORY SAMPLE PREPARATION

1. Drying

A field-collected soil sample arriving at the laboratory is usually assigned a laboratory number and then air-dried, either in an ambient air circulating cabinet or an air circulating oven, the air possibly being warmed slightly above the laboratory ambient temperature. The drying process should be done as promptly and rapidly as possible to minimize microbial activity (mineralization).

The time required to bring a soil sample to an air-dried condition will be determined by its moisture, texture, and organic matter contents. Soils high in clay and/or organic matter content will take a considerably longer time to bring to an air-dried condition than sandy-textured soils.

Drying can be facilitated by exposing as much surface of the soil to circulating air as possible and by elevating the drying temperature. Temperature should not exceed 38°C (100°F), as significant changes in the physio-chemical properties of the soil can occur at elevated drying temperatures.

The drying of some types of soils will result in a significant release or fixation of potassium (K) (Goulding, 1987; Sparks, 1987); therefore, for some determinations, the arriving soil sample may be assayed as received without removing field moisture (Goulding, 1987; Bates, 1993). In addition, the determination of the micronutrients, copper (Cu), iron (Fe), manganese (Mn), and zinc (Zn) can be affected by the drying process (Kahn and Soltanpour, 1978; Shuman, 1980).

The moisture content of an air-dried soil will be determined by the physio-chemical properties of the soil and the relative humidity of air surrounding the sample. This variability will have little effect on most soil analysis procedures, the minimal effect being when the soil aliquot is by volume measurement rather than by weight.

2. Crushing/Grinding

Following drying, the soil sample is crushed, either by hand using a pestle and mortar or by using a mechanical device, and then passed through a 10-mesh (2-mm) screen. A typical mechanical grinding and sieving device is shown in Figure 2.1. This

FIGURE 2.1 Soil grinding and sieving device (Custom Laboratory Equipment, Orange City, FL).

preparation procedure can contaminate a soil sample either from the composition of contacting surfaces or from deposition of dust and/or a previous sample residue.

Although crushing and sieving can be a mixing process, sample size reduction may be necessary and care must be exercised to ensure that the sample is thoroughly mixed before dividing.

Grinding can have an effect on some elemental determinations, as has been discussed by Kahn (1979) for the determination of copper (Cu), iron (Fe), and zinc (Zn).

2.3 SAMPLE ALIQUOT DETERMINATION

1. Weighing versus Scooping

Soil aliquot transfer to a saturation or extracting vessel is commonly done by weighing. But in order to speed the transfer of a prepared soil sample, the use of scoops to either obtain an estimated weight of sample or a volume of soil is commonly practiced in many soil testing laboratories. The use of volume as the measurement for aliquot amount has been recommended by Mehlich (1972, 1973). Bates (1993) has discussed weight versus volume measurement considerations, and van Lierop (1989) has compared weight versus volume measurement of soil aliquots on accuracy of the assay result.

In this handbook, both volume and estimated-weight scoops are used to obtain the soil aliquot for many determinations as well as determinations based on weighed samples. In most instances, the method most commonly associated with that procedure is specified.

2. Estimated Weight Scoops

Scoop size is based on an assumed "average" volume-weight of prepared sample, air-dried, and 10-mesh (2-mm) soil. The typical soil prepared for analysis, as described in this handbook, has an assumed weight-to-volume ratio of 1.18 for silt loam and clay textured soils and 1.25 for sandy soils. Therefore, those soil test procedures adapted to a particular textured soil will designate scoop volumes that match the assumed weight-to-volume ratio:

Silt Loam and Clay Textured Soils		Sandy Soils	
weight g	scoop size cm^3	weight g	scoop size cm^3
2.5	1.70	5.0	4.0
5.0	4.25		
10.0	8.50		

Scoops are of a fixed volume and do not necessarily yield an estimated or assumed weight. However, when the volume weight of a soil sample is known, a specific volume of that soil can be scooped to give an estimated weight.

In most instances, a dual system of weighed and/or volume-measured samples is presented. This rationale is necessary in cases in which the original method specified a weight of sample or volume of known or assumed specific weight. The reader may refer to Mehlich (1972, 1973) and van Lierop (1981, 1989) for additional information on volume-weight considerations and to Peck (1998) for more details on scoop design and use.

3. NCR-13 Scoops

The design of the NCR-13 scoops commonly used by soil testing laboratories in the north central region of the United States has been described by Peck (1998). As shown in Figure 2.2, the specifications are:

NCR-13 Standard Soil Scoop Specifications (Manufactured from Stainless Steel).

Scoop Size[a]	Scoop Capacity	Outside Diameter	Inside Diameter	Inside Diameter
g	cc	inch	inch	inch
1	0.85	5/8	1/2	17/64
2	1.70	3/4	5/8	22/64
5	4.25	1	7/8	28/64
10	8.50	1-1/4	1-1/8	34/64

[a]Grams of soil in terms of the "typical" soil (defined as a silt loam texture with 1.25% organic matter crushed to pass a 10-mesh screen; bulk density of crushed "typical" soil approximates 1.18 compared with 1.32 for "undisturbed" soil) weighing 2,000,000 pounds per acre in the top 6-2/3 inch layer.

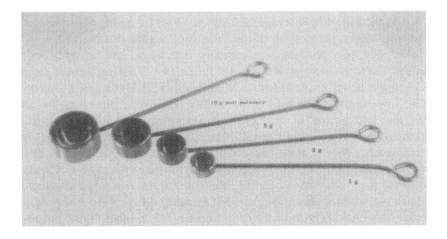

FIGURE 2.2 NCR-13 standard soil scoops.

4. Procedure for Using a Soil Scoop

1. Stir the crushed and screened sample with a spatula to loosen soil prior to measuring.
2. Dip into the center of the soil sample with the soil scoop, filling it heaping full without pushing against the side of the soil container.
3. Hold the scoop firmly and tap the handle three times with a spatula from a distance of two or three inches.
4. Hold the spatula blade perpendicular to the top of the scoop and strike off excess soil. A flat spatula blade may be replaced by a round rod, which keeps from scarring the leveled surface.
5. Empty the scoop into an appropriate extraction vessel.

Since an accurate measure for a scooped sample is essential, scoop design is a very important factor. The diameter of the scoop should be twice its height to ensure the most efficient packing density in the scoop.

Variance among repeated scoopings of a soil sample will be within 2–3% of the same volume or estimated weight. In general, scooping of soil samples has been found to give results comparable to weighed samples in repeated analyses of the same soil sample.

2.4 LABORATORY FACTORS

1. Extraction Reagents

Many of the extraction reagents currently in use today (Jones, 1990, 1998) reflect the history of their development and use. Extraction reagents that were developed for specific applications in the 1940s and 1950s are now considered standard procedures for the determination of one or more nutrient elements. For example, Bray P1 for P determination (Bray and Kurtz, 1945) in acid to neutral pH soils, Olsen (Olsen et al., 1954) for alkaline soils, and neutral normal ammonium acetate for the determination of K, Ca, and Mg (Schollenberger and Simon, 1945) for both acid and alkaline soils were, and still are, the methods of choice in many laboratories.

The first two procedures, commonly referred to as "universal extraction reagents," were the Morgan (Morgan, 1941; Lunt et al., 1950) procedure for use on a wide range of soil types, and Mehlich No. 1 (Mehlich, 1953) for application on sandy, acidic, low organic matter soils of the southeastern coastal plain region of the U.S. With the introduction of multielement analyzers, such as various forms of AutoAnalyzers (Watson and Isaac, 1990), and inductively coupled plasma emission spectrometers (Watson and Isaac, 1990; Soltanpour et al., 1998), one extraction reagent for the determination of many elements, major elements (P, K, Ca, and Mg), and of the micronutrients (B, Cu, Fe, Mn, and Zn) resulted in the development of Mehlich No. 3 extractant (Mehlich, 1984) for many different types of soils, and for alkaline soils, AB-DTPA extractant (Soltanpour, 1991). Jones (1990) has reviewed the development and application of the universal extraction reagents. Recently, the adaptation of the 0.01 M $CaCl_2$ extraction procedure for multielement determination has been proposed by Houba and his colleagues (Houba et al., 1990).

2. Extraction Procedure

Extraction procedure parameters, such as the shape and size of the extraction vessel (Wheaton bottle versus Erlenmeyer flask), shaking speed, and temperature, can have a significant effect on the extraction of P and K from a soil by most of the commonly used extraction procedures (Eliason, 1998). Therefore, control of these factors is essential if the obtained assay result is to be reliable.

3. Reagents and Standards

Careful preparation, storage, and use of reagents and standards are critical in order to successfully carry out the procedures described in this handbook. One frequently overlooked factor that can affect an analytical result is the pH of the extraction reagent. For those assay procedures given in this handbook, the user should be aware of these effects and carefully follow the procedures as given without modification to ensure that reliable assay results are obtained.

2.5 LONG-TERM STORAGE

Since there have been no specific studies that would define the required storage conditions and identify those changes that will occur from long-term storage of laboratory prepared soil samples, it has been suggested that soils are best able to maintain their original integrity when stored in an air-dried condition at low humidity and just above the freezing temperature. However, some result data indicate that some soil parameters may change during such storage conditions (Bates, 1993; Houba and Novozamsky, 1998; Brown, 1999).

2.6 REFERENCES

Anon. 1994. Soil sample preparation, In: *Plant, Soil, and Water Reference Methods for the Western Region.* A Western Regional Extension Publication WREP 125. R. G. Galvak, D. A. Horneck, and R. O. Miller (eds.). Fairbanks, AK: University of Alaska, 17.

Anon. 1999. Soil sampling strategies key. *Ag. Consultant* 55:17–20.

Bates, T. E. 1993. Soil handling and preparation, In: *Soil Sampling and Methods of Analysis,* M. R. Carter (ed.). Boca Raton, FL: CRC Press, 19–24.

Bray, R. H. and L. T. Kurtz. 1945. Determination of total, organic, and available forms of phosphorus in soils. *Soil Sci.* 59:39–45.

Brown, A. J. 1999. Soil sampling and sample handling for chemical analysis, In: *Soil Analysis: An Interpretation Manual.* K. I. Peverill, L. A. Sparrow, and D. J. Reuter (eds.), Collingwood, Australia: CSIRO Publishing, 35–53.

Crépin, J. and R. L. Johnson. 1993. Soil sampling for environmental assessment, In: *Soil Sampling and Methods of Analysis,* M. R. Carter (ed.). Boca Raton, FL: CRC Press, 5–10.

Eliason, R. 1998. Laboratory factors of importance to soil extraction, In: *Recommended Chemical Soil Test Procedures for the North Central Region,* North Central Regional Publication No. 221 (revised). J. R. Brown (ed.). Columbia, MO: Missouri Agricultural Experiment Station SB1001, University of Missouri, 11–12.

Gelderman, R. H. and A. P. Mallarino. 1998. Soil sample preparation, In: *Recommended Chemical Soil Test Procedures for the North Central Region,* North Central Regional

Publication No. 221 (revised). J. R. Brown (ed.). Columbia, MO: Missouri Agricultural Experiment Station SB1001, University of Missouri, 5–6.

Goulding, K. W. T. 1987. Potassium fixation and release. *Proceedings of Colloquium International Potash Institute* 20:137–154.

Hoskins, B. and D. Ross. 1995. Soil sample preparation and extraction, In: *Recommended Soil Testing Procedures for the Northeastern United States,* 2nd ed. Northeast Regional Publication Bulletin 493. J. T. Sims and A. M. Wolf (eds.). Newark, DE: Agricultural Experiment Station, University of Delaware, 10–15.

Houba, J. V. G. and I. Novozamsky. 1998. Influence of storage time and temperature of air-dried soils on pH and extractable nutrients using 0.01 M $CaCl_2$. *Fresenius J. Anal. Chem.* 360:362–365.

Houba, J. V. G., I. Novozamsky, Th. M. Lexmond, and J. J. van der Lee. 1990. Applicability of 0.01 M $CaCl_2$ as a single extractant for the assessment of the nutrient status of soils and other diagnostic purposes. *Commun. Soil Sci. Plant Anal.* 21:2281–2290.

James, D. W. and K. L. Wells. 1990. Soil sample collection and handling: Techniques based on source and degree of field variability, In: *Soil Testing and Plant Analysis,* R. L. Westerman (ed.). 3rd ed. SSSA Book Series No. 3. Madison, WI: Soil Science Society of America, 25–44.

Jones, Jr., J. B. 1990. Universal soil extractants: Their composition and use. *Commun. Soil Sci. Plant Anal.* 21:1091–1101.

Jones, Jr., J. B. 1998. Soil test methods: Past, present, and future. *Commun. Soil Sci. Plant Anal.* 29:1543–1552.

Kahn, A. 1979. Distribution of DTPA-extractable Fe, Zn, and Cu in soil particle-size fractions. *Commun. Soil Sci. Plant Anal.* 10:1211–1218.

Kahn, A. and P. N. Soltanpour. 1978. Effect of wetting and drying on DTPA-extractable Fe, Zn, Mn, and Cu in soils. *Commun. Soil Sci. Plant Anal.* 9:193–202.

Lunt, H. A., C. L. W. Swanson, and H. G. H. Jacobson. 1950. *The Morgan Soil Testing System.* Connecticut Agricultural Experiment Station (New Haven) Bulletin 541.

Mehlich, A. 1953. *Determination of P, Ca, Mg, K, Na, and NH_4.* Raleigh, NC: Mimeo., North Carolina Soil Testing Division.

Mehlich, A. 1972. Uniformity of expressing soil test results: A case of calculating results on a volume basis. *Commun. Soil Sci. Plant Anal.* 3:417–424.

Mehlich, A. 1973. Uniformity of soil test results as influenced by volume weight. *Commun. Soil Sci. Plant Anal.* 4:475–486.

Mehlich, A. 1984. Mehlich 3 soil test extractant: A modification of Mehlich 2 extractant. *Commun. Soil Sci. Plant Anal.* 15:1409–1416.

Morgan, M. F. 1941. *Chemical Diagnosis by the Universal Soil Testing System.* Connecticut Agricultural Experiment Station Bulletin 450.

Nowak, P. 1998. Agriculture and change: The promises and pitfalls of precision. *Commun. Soil Sci. Plant Anal.* 29:1537–1541.

Olsen, S. R., C. V. Cole, F. S. Watanabe, and L. A. Dean. 1954. *Estimation of Available Phosphorus in Soils by Extraction with Sodium Bicarbonate.* USDA Circular 939. U.S. Government Printing Office, Washington, D.C.

Peck, T. R. 1998. Standard soil scoop, In: *Recommended Chemical Soil Test Procedures for the North Central Region,* North Central Regional Publication No. 221 (revised). J. R. Brown (ed.). Columbia, MO: Missouri Agricultural Experiment Station SB 1001, University of Missouri, 7–9.

Peck, T. R. and R. Beck. 1998. Soil sampling and soil availability, past, and present. *Commun. Soil Sci. Plant Anal.* 29:1425–1428.

Peterson, R. G. and L. D. Calvin. 1996. Sampling, In: *Methods of Soil Analysis, Part 3, Chemical Methods,* SSSA Book Series No. 5 R. L. Sparks (ed.). Madison, WI: Soil Science Society of America, 1–17.

Radojevic, M. and V. N. Bashkin. 1999. *Practical Environmental Analysis.* Cambridge: The Royal Society of Chemistry.

Reid, K. (ed.). 1998. *Soil Fertility Handbook.* Ministry of Agriculture, Food and Rural Affairs. Toronto: Queen's Printer for Ontario.

Sabbe, W. E. and D. B. Marx. 1987. Soil sampling: Spatial and temporal variability, In: *Soil Testing: Sampling, Correlation, Calibration, and Interpretation,* SSSA Special Publication No. 21. J. R. Brown (ed.). Madison, WI: Soil Science Society of America, 1–14.

Schollenberger, C. J. and R. H. Simon. 1945. Determination of exchange capacity and exchangeable bases in soils—ammonium acetate method. *Soil Sci.* 59:13–24.

Schnug, E., K. Panten, and S. Haneklaus. 1998. Sampling and nutrient recommendations—the future. *Commun. Soil Sci. Plant Anal.* 29:1455–1462.

Shuman, L. M. 1980. Effects on soil temperature, moisture, and air-drying on extractable manganese, iron, copper, and zinc. *Soil Sci.* 130:336–343.

Soltanpour, P. N. 1991. Determination of nutrient element availability and elemental toxicity by the AB-DTPA soils test and ICPS. *Adv. Soil Sci.* 16:165–190.

Soltanpour, P. N., G. W. Johnson, S.M. Workman, J. B. Jones, Jr., and R. O. Miller. 1998. Advances in ICP emission and ICP mass spectrometry. *Adv. Agron.* 64:28–113.

Sparks, D. L. 1987. Potassium soil dynamics in soils. *Adv. Soil Sci.* 6:1–63.

van Lierop, W. 1981. Laboratory determination of field bulk density for improving fertilizer recommendations of organic soils. *Can. J. Soil Sci.* 61:475–482.

van Lierop, W. 1989. Effect of assumptions on accuracy of analytical results and liming recommendations when testing a volume or weight of soil. *Commun. Soil Sci. Plant Anal.* 20:121–137.

Watson, M. E. and R. A. Isaac. 1990. Analytical instruments for soil and plant analysis, In: *Soil Testing and Plant Analysis,* 3rd ed., SSSA Book Series No. 3, R.L. Westerman (ed.). Madison, WI: Soil Science Society of America, 691–740.

Wright, N. A. 1998. Soil fertility variograms from "True Point Sampling™" on 20.0, 0.9, and 0.1 meter grids in two fields. *Commun. Soil Sci. Plant Anal.* 29:1649–1666.

Soil Sampling Devices Suppliers:

Clements Associates, 1992 Hunter Ave., Newton, IA 50208-8652 (800-247-6630; fax: 515-792-1361)

Concord Environmental Equipment, RR 1, Box 78, Hawley, MN 56449-9739 (218-937-5100; fax: 218-937-5101)

Geophyta, 2685 County Road 254, Vickery, OH 43464-9775 (419-547-8538)

Linco Equipment, Inc., I-39 and US 24W, El Paso, IL 61738 (309-527-6455; fax: 309-527-660)

Oakfield Apparatus, P.O. Box 65, Oakfield, WI 53065-0065 (414-583-4114; fax: 414-583-4166)

Western Ag Innovations, 217 Badger Court, Saskatoon, SK, Canada S7N 2X2 (306-249-3237; fax: 306-249-3237)

3 Soil pH, and Exchangeable Acidity and Aluminum

The pH of a soil significiantly affects plant growth, primarily due to the change in availability of both the essential elements, such as phosphorus (P) and most of the micronutrients, copper (Cu), iron (Fe), manganese (Mn), molybdenum (Mo), and zinc (Zn), as well as nonessential elements, such as aluminum (Al), that can be toxic to plants at elevated concentrations (Woodruff, 1967; Black, 1993; Slattery et al., 1999). The activities of microbial populations are also affected by pH as well as the activites of some types of pest chemicals applied to soils.

The soil pH is a measure of the hydronium ion (H_3O^+, or more commonly the H^+) activity and is defined as the negative logarithm (base 10) of the H^+ activity (moles per liter) in the soil solution (Peach, 1965; Coleman and Thomas, 1967). By definition, soils having a pH less than 7.0 are generally defined as "acidic," when above pH 7.0 as "alkaline," and at pH 7.0, "neutral." Two classification systems based on water pH are:

Category	pH	Category	pH
very acid	4.5–5.5	acid	< 4.5
acid	5.6–6.0	weakly acid	4.5–6.5
slightly acid	6.1–6.8	neutral	6.6–7.5
neutral	6.9–7.6	weakly basic	7.6–9.5
alkaline	7.7–8.3	basic	>9.5

Generally, more descriptive terms are used to define pH in terms of its probable effect on the soil itself as well as plant growth and development (Woodruff, 1967).

Although seemingly a simple measurement that can be made using chemical dyes (Woodruff, 1961), but more commonly by the use of a pH meter, an accurate measurement of pH can be difficult to make depending on soil characteristics and the technique used to make the measurement (Schofield and Taylor, 1955; McLean, 1973; Conyers and Darey, 1988; van Lierop, 1990). When using a pH meter, the soil pH can be determined in a soil:water slurry of various ratios, normally 1:1 or 1:2 (Anon., 1994; Watson and Brown, 1998), or in a soil slurry of either 0.01 M calcium chloride ($CaCl_2 \cdot 2H_2O$) (Plank, 1992a) or 1N potassium chloride (KCl) (Plank, 1992b). A comparison of the pH measurement in 0.01 M $CaCl_2 \cdot 2H_2O$ and 1N KCl is shown in Figure 3.1 (Fotyma et al., 1998).

The positioning of the pH meter electrodes as well as the stirring of the slurry can significantly influence the pH determination. An excellent reference on the

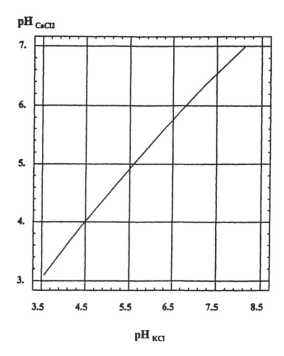

FIGURE 3.1 Regression line for the relation between pH-CaCl$_2$ pH-KCl (Fotyma, 1998).

measurement of a soil pH and the maintenance of pH electrodes is given by Sumner (1994). The basic chemistry of soil acidity has been described by Peach (1965), Coleman and Thomas (1967), Hendershot et al. (1993), and Thomas (1996). General instructions on the measurement of pH and factors influencing such measurements have been described by Meier, et al. (1989), Sartoretto (1991), and Sims and Eckert (1995). In an agreement between the Association of Official Analytical Chemists (AOAC) and the Soil Science Society of America (SSSA) described by Kalra (1996), a standardized procedure for the determination of pH has been published (Kalra, 1995).

For an interprepation of a soil pH determination, refer to Pearson and Adams (1967), Alley and Zelazny, 1987; Black (1993), Anon. (1996), Maynard and Hochmuth (1997), and Reid (1998). Data on soils having a water pH of less than 6.0 in North America for 1997 has been prepared by the Potash & Phosphate Institute (Anon., 1998).

3.1 pH IN WATER

1. Principle of the Method

1.1. This procedure is used to determine the pH of a soil in a water suspension. pH is defined as "the negative logarithm to the base 10 of the H$^+$ ion concentration, or the logarithm of the reciprocal of the H^+ ion concentration in the soil solution." Since the pH is logarithmic, the H^+ ion concentration in solution increases 10 times when the pH is lowered one unit.

2. Range and Sensitivity

2.1. A pH meter equipped with glass hydrogen and calomel reference electrodes is used to measure the pH in water for soils between pH 3.5 and pH 9.0.
2.2. The sensitivity of the measurement will depend on the pH meter and electrodes. For routine soil testing, it is only necessary to read the pH to the 0.1 unit.

3. Interferences

3.1. Hydrogen (H) ions may be displaced from the exchange sites, and the presence of other ions causes additional H^+ ions to form in solution. This interference results in a lower pH.
3.2. Carbon dioxide (CO_2) from the atmosphere or soil air dissolved in water, forming carbonic acid (H_2CO_3), can lower the pH markedly. Only in soils that have a pH considerably above 7.0, i.e., a very low H^+ ion concentration, does the CO_2 concentration of the air have an appreciable effect on the pH measurement.

4. Precision and Accuracy

4.1. Random variation of 0.1 to 0.2 pH units is allowable in replicate determinations, and this variation can be expected from one laboratory to another.
4.2. Scratching of the glass electrode can result in erratic performance. Likewise, the reference electrode with restricted flow of the filling solution or low filling solution level can result in unstable readings.
4.3. Dehydrated electrodes give erratic readings. Follow the electrode manufacturer's instructions in keeping the electrodes hydrated.

5. Apparatus

5.1. No. 10 (2-mm opening) sieve.
5.2. 4.25-cm^3 volumetric scoop for soils to be used for SMP buffer pH determination (see Chapter 4) and a 10-cm^3 volumetric scoop for soils to be used for Adams-Evans buffer pH determination (see Chapter 4).
5.3. 50-mL cup (glass, plastic, or waxed paper of similar size).
5.4. Pipettes, 5-mL and 10-mL capacity.
5.5. Stirring apparatus (mechanical shaker, stirrer, or glass rod).
5.6. pH meter with reproducibility to at least 0.05 pH units, equipped with a glass electrode paired with a calomel reference electrode or combination electrode.
5.7. Glassware and dispensing apparatus for the preparation and dispensing of buffer solutions.
5.8. Analytical balance.

6. Reagents (use reagent grade chemicals and pure water)

6.1. *pH 10.0 Buffer Solution:* Commercially prepared pH buffer, or weigh 1.956 g sodium hydroxide (NaOH) and 3.092 g boric acid (H_3BO_3) into a 1,000-mL volumetric flask and bring to volume with water.

6.2. *pH 7.0 Buffer Solution:* Commercially prepared pH buffer, or weigh 3.3910 g citric acid ($C_6H_8O_7$) and 23.3844 g disodium phosphate ($Na_2HPO_4 \cdot 12H_2O$) into a 1,000-mL volumetric flask and bring to volume with water.

6.3. *pH 4.0 Buffer Solution:* Commercially prepared pH buffer, or weigh 11.8060 g citric acid ($C_6H_8O_7$) and 10.9468 g disodium phosphate ($Na_2HPO_4 \cdot 12H_2O$) into a 1,000-mL volumetric flask and bring to volume with water.

7. Procedure

7.1. *Determination:* Weigh 5 g, or scoop 4.25 cm^3 for a following SMP buffer determination, or scoop 10 cm^3 for a following Adams-Evans buffer (see Chapter 4) determination <10-mesh (2-mm) soil into a cup (see 5.3). Pipette 5 mL or 10 mL (for 4.25-cm^3 or 10-cm^3 soil, respectively) water into the cup and stir for five seconds. Let stand for 10 minutes. Calibrate the pH meter according to the instructions supplied with the specific meter (see 8.1). Stir the soil and water slurry. Lower the electrodes into the soil-water slurry so that the electrode tips are at the soil-water interface. While stirring the soil water slurry, read the pH, and record to the nearest tenth of a pH unit. Erratic movement of the pH meter dial or number may be due to faulty-operating electrodes or lack of sufficient junction potential (Sumner, 1994).

7.2. Save the sample for the determination of the buffer pH if the soil water pH is less than 6.0 (see Chapter 4).

8. Calibration of pH Meter

8.1. Calibrate the pH meter using buffer solutions pH 7.0 and pH 4.0 (see 6.2 and 6.3) or pH 10.0 and pH 7.0 (see 6.1 and 6.2) that will bracket the range for the unknown soils to be tested following the meter's instruction manual.

9. Calculation

9.1. The result is reported as water pH (pH_w).

10. Effects of Storage

10.1. Air-dry soils may be stored for several months in closed containers without affecting the pH_w measurement.

10.2. If the pH meter and electrodes are not to be used for an extended period of time, the instructions for storage given by the instrument manufacturer should be followed.

11. Interpretation

11.1. For an interprepation of a soil pH determination, refer to Pearson and Adams (1967), Black (1993), Anon. (1996), Maynard and Hochmuth (1997), and Reid (1998).

3.2 SOIL pH IN 0.01 M CALCIUM CHLORIDE

1. Principle of the Method

1.1. This method estimates the activity of hydrogen (H^+) ions in a soil suspension in the presence of 0.01 M calcium chloride ($CaCl_2$) to approximate a constant ionic strength for all soils, regardless of past management, mineralogical composition, and natural fertility level.

1.2. The use of 0.01 M $CaCl_2$ in soil pH measurement was proposed by Schofield and Taylor (1955) and has been recently described by Fotyma et al. (1988). Peach (1965) summarized the advantages of using 0.01 M $CaCl_2$ for measuring soil pH. The merits of determining soil pH in a constant salt level are also discussed by Woodruff (1967), McLean (1973), and Conyers and Darey (1988).

2. Range and Sensitivity

2.1. A pH meter equipped with glass hydrogen and calomel reference electrodes is used to measure the pH in 0.01 M $CaCl_2$ for soils that range in pH from 2.5 to 8.0, which would include most soils.

2.2. The sensitivity will depend on the pH meter. For routine soil testing, it is only necessary to read the pH to the 0.1 unit.

2.3. The pH in 0.01 M $CaCl_2$ may be estimated using the bromocresol purple dye procedure described by Woodruff (1961).

3. Interferences

3.1. The main advantage of the measurement of soil pH in 0.01 M $CaCl_2$ is the elimination of interferences from suspension effects and from variable salt contents, such as fertilizer residues.

4. Precision and Accuracy

4.1. Temperate region soil pH values in 0.01 M $CaCl_2$ are lower in magnitude (higher H^+ ion concentration) and less variable than those made in water, due to the release of H^+ ions from exchange sites by calcium (Ca^{2+}) ions.

4.2. Scratching of the glass electrode can result in erratic performance. Likewise, the reference electrode with restricted flow of the filling solution or low filling solution level can result in unstable readings.

4.3. Dehydrated electrodes give erratic readings. Follow the electrode manufacturer's instructions in keeping the electrodes hydrated.

5. Apparatus

5.1. No. 10 (2-mm opening) sieve.
5.2. 4.25-cm^3 scoop, volumetric.
5.3. 50 mL-cup, (glass, plastic, or waxed paper of similar size).
5.4. Pipette, 5-mL capacity.

5.5. Stirring apparatus (mechanical shaker, stirrer or glass rod).

5.6. pH meter with reproducibility to at least 0.05 pH unit, and a glass electrode paired with a calomel reference electrode.

5.7. Glassware and dispensing apparatus for the preparation and dispensing of 0.01 M CaCl$_2$ and buffer solutions.

5.8. Dropping bottle, 30- or 60-mL capacity (see alternate procedure 7.2 below).

5.9. Analytical balance.

6. Reagents (use reagent grade chemicals and pure water)

6.1. *0.01 M Calcium Chloride:* Weigh 1.47 g calcium chloride dihydrate (CaCl$_2$·2H$_2$O) into a 1,000-mL volumetric flask and bring to volume with water.

6.2. *pH 10.0 Buffer Solution:* Commercially prepared pH buffer, or weigh 1.756 g sodium hydroxide (NaOH) and 3.092 g boric acid (H$_3$BO$_3$) into a 1,000-mL volumetric flask and bring to volume with water.

6.3. *pH 7.0 Buffer Solution:* Commercially prepared pH buffer, or weigh 3.391 g citric acid (C$_6$H$_8$O$_7$) and 23.3844 g disodium phosphate (Na$_2$HPO$_4$·12H$_2$O) into a 1,000-mL volumetric flask and bring to volume with water.

6.4. *pH 4.0 Buffer Solution:* Commercially prepared buffer, or weigh 11.806 g citric acid (C$_6$H$_8$O$_7$) and 10.9468 g disodium phosphate (Na$_2$HPO$_4$·12H$_2$O) into a 1,000-mL volumetric flask and bring to volume with water.

6.5. *1.0 M Calcium Chloride:* Weigh 147 g calcium chloride dihydrate (CaCl$_2$·2H$_2$O) into a 1,000 mL volumetric flask and bring to volume with water.

7. Procedure

7.1. *Determination:* Weigh 5 g air-dry, or scoop 4.25 cm^3 < 10-mesh (2-mm) soil into a 50 mL cup (see 5.3). Pipette 5 mL 0.01 M CaCl$_2$ solution (see 6.1) into the cup and stir for 30 minutes on a mechanical stirrer or shaker (or stir periodically with a glass rod for a period of 30 minutes). Calibrate the pH meter according to the instructions supplied with the specific meter (see 8.1). Stir the soil and 0.01 M CaCl$_2$ slurry. Lower the electrodes into the soil-0.01 M CaCl$_2$ slurry so that the electrode tips are at the soil-water interface. While stirring the soil water slurry, read the pH, and record to the nearest tenth of a pH unit.

7.2. For laboratories desiring both soil pH in water and 0.01 M CaCl$_2$, 5 mL water can be substituted for the 5 mL 0.01 M CaCl$_2$ as given in 7.1. After the water pH is determined, add one drop of 1 M CaCl$_2$ (see 6.5) to the soil-water suspension, stir or shake for 30 minutes, and then read the pH of the suspension and designate as pH$_{CaCl2}$.

8. Calibration of pH Meter

8.1. Calibrate the pH meter using buffer solutions pH 7.0 and pH 4.0 (see 6.3 and 6.4) or pH 10.0 and pH 7.0 (see 6.2 and 6.3) that will bracket the range for the unknown soils to be tested following the meter's instruction manual.

9. Calculation

9.1. The result is reported as pH$_{CaCl2}$.

10. Effects of Storage

10.1. Air-dry soils may be stored several months in closed containers without affecting the pH_{CaCl2} measurement (Houba and Novozamsky, 1998).
10.2. If the pH meter and electrodes are not to be used for an extended period of time, the instructions for storage given by the instrument manufacturer should be followed.

11. Interpretation

11.1. For an interprepation of a soil pH determination, refer to Pearson and Adams (1967), Black (1993), Anon. (1996), Maynard and Hochmuth (1997), Reid (1998), and Fotyma et al. (1998).

3.3 SOIL pH IN 1 N POTASSIUM CHLORIDE

1. Principle of the Method

1.1. This procedure is used to determine the pH of a soil in 1 N potassium chloride (KCl). pH is defined as "the negative logarithm to the base 10 of the H^+ ion concentration, or the logarithm of the reciprocal of the H^+ ion concentration in the soil solution." Since the pH is logarithmic, the H^+ ion concentration in solution increases 10 times when the pH is lowered one unit.

2. Range and Sensitivity

2.1. A pH meter equipped with glass hydrogen and calomel reference electrodes is used to measure the pH in 1 N KCl for soils between pH 3.5 and pH 9.0.
2.2. The sensitivity will depend on the pH meter. For routine soil testing, it is only necessary to read the pH to the 0.1 unit.

3. Interferences

3.1. None.

4. Precision and Accuracy

4.1. Random variation of 0.1 to 0.2 pH unit is allowable in replicate determinations, and this variation can be expected from one laboratory to another.
4.2. Scratching of the glass electrode can result in erratic performance. Likewise, the reference electrode with restricted flow of the filling solution or low filling solution level can result in unstable readings.
4.3. Dehydrated electrodes give erratic readings. Follow the electrode manufacturer's instructions in keeping the electrodes hydrated.

5. Apparatus

5.1. No. 10 (2-mm opening) sieve.
5.2. 4.25-cm^3 volumetric scoop.

5.3. 50-mL cup (glass, plastic, or waxed paper of similar size).
5.4. Pipettes, 5-mL and 10-mL capacity.
5.5. Stirring apparatus (mechanical shaker, stirrer, or glass rod).
5.6. pH meter with reproducibility to at least 0.05 pH unit, and a glass electrode paired with a calomel reference electrode.
5.7. Glassware and dispensing apparatus for the preparation and dispensing of buffer solutions.
5.8. Analytical balance.

6. Reagents (use reagent grade chemicals and pure water)

6.1. *1 N Potassium Chloride:* Weight 74.56 g potassium chloride (KCl) into a 1,000-mL volumetric flask and bring to volume with water.
6.2. *pH 7.0 Buffer Solution:* Commercially prepared pH buffer, or weigh 3.3910 g citric acid ($C_6H_8O_7$) and 23.3844 g disodium phosphate ($Na_2HPO_4 \cdot 12H_2O$) into a 1,000 mL volumetric flask and bring to volume with water.
6.3. *pH 4.0 Buffer Solution:* Commercially prepared pH buffer, or weigh 11.8060 g citric acid ($C_6H_8O_7$) and 10.9468 g disodium phosphate ($Na_2HPO_4 \cdot 12H_2O$) into a 1,000-mL volumetric flask and bring to volume with water.

7. Procedure

7.1. *Determination:* Weigh 5 g, or scoop 4.25 cm^3 <10-mesh (2-mm) soil into a cup (see 5.3), pipette 5 mL 1 N KCl (see 6.1) into the cup, and stir for five seconds. Let it stand for 10 minutes. Calibrate the pH meter according to the instructions supplied with the specific meter (see 8.1). Stir the soil and water slurry. Lower the electrodes into the soil-water slurry so that the electrode tips are at the soil-water interface. While stirring the soil water slurry, read the pH and record to the nearest tenth of a pH unit.

8. Calibration of pH Meter

8.1. Calibrate the pH meter using buffer solutions pH 7.0 and pH 4.0 (see 6.2 and 6.3) that will bracket the range for the unknown soils to be tested following the meter's instruction manual.

9. Calculation

9.1. The result is reported as pH_{KCl}.

10. Effects of Storage

10.1. Air-dry soils may be stored several months in closed containers without affecting the pH_{KCl} measurement.
10.2. If the pH meter and electrodes are not to be used for an extended period of time, the instructions for storage given by the instrument manufacturer should be followed.

11. Interpretation

11.1. For an interprepation of a soil pH determination, refer to Pearson and Adams (1967), Black (1993), Anon. (1996), Maynard and Hochmuth (1997), and Reid (1998).

3.4 DETERMINATION OF EXCHANGEABLE ACIDITY USING BARIUM CHLORIDE-TEA BUFFER

1. Principle of the Method

1.1. This method measures the acidity that is exchangeable by the barium chloride ($BaCl_2$)-TEA extractant that is buffered at pH 8.2. Thus the exchangeable acidity measured is comprised of exchangeable aluminum (Al) and any hydrogen (H) that will dissociate when the soil is brought to a pH of 8.2 (potential acidity) (Sumner, 1992a). It is also a measure of the variable charge developed between the soil pH and pH 8.2.

1.2. This method was developed by Mehlich (1953) and is a modification of a previous Mehlich method (1939).

2. Range and Sensitivity

2.1. By varying the quantity of soil used, this method can be used on all soils to measure exchangeable acidity.

2.2. Because the endpoint is determined spectrophotometrically, some variation between operators can be expected.

3. Interferences

3.1. Few problems are experienced.

4. Precision and Accuracy

4.1. Exchangeable acidity can be determined within 0.1 meq per 100 g.

5. Apparatus

5.1. Beakers, 100-mL glass.
5.2. Buchner funnel (5.5 cm) and vacuum flask.
5.3. Whatman No. 42 filter paper, 5.5 cm.

6. Reagents (use reagent grade chemicals and pure water)

6.1. *Buffer Solution:* Adjust 0.5 N barium chloride dihydrate ($BaCl_2 \cdot 2H_2O$) (61.07 g L^{-1}) and 0.2 N triethanolamine [$N(CH_2CH_2OH)_3$] (29.8 g L^{-1}) to pH 8.2 with hydrochloric acid (HCl). Protect from carbon dioxide (CO_2) contamination by attaching a tube containing soda lime to the air intake.

6.2. *Replacing Solution:* Combine 0.5 N barium chloride dihydrate (61.07 g BaCl$_2$·2H$_2$O in 1,000 mL water) with 0.4 mL Buffer Solution (see 6.1) per 1,000 mL. Protect from carbon dioxide (CO$_2$) as with the Buffer Solution (see 6.1).

6.3. *Hydrochloric Acid (HCl):* Approximately 0.2N standard.

6.4. *Bromocresol Green:* 0.1% aqueous solution.

6.5. *Mixed Indicator:* Weigh 1.250 g methyl red and 0.825 g methylene blue in 1,000 mL 90% ethanol.

7. Procedure

7.1. Scoop 10 cm^3 air dry <10-mesh soil into 100 mL beaker (use 5 cm^3 for very acid soils), add 25 mL Buffer Solution (see 6.1), mix well, and allow to stand for 1 hour. Transfer mixture to Buchner filtration system and add a further three aliquots (25 mL) Buffer Solution (see 6.1). Continue with 100 mL Replacing Solution (see 6.2) for a total of 200 mL.

7.2. Mix 100 mL Buffer Solution (see 6.1) with 100 mL Replacing Solution (see 6.2) to serve as a blank. Add two drops bromocresol green (see 6.4) and 10 drops mixed indicator (see 6.5). Titrate with HCl (see 6.3) to a green to purple endpoint. Follow same method for soil filtrates.

8. Calculations

$$\text{exchangeable acidity} = \frac{\text{mL HCl for blank} - \text{mL HCl for soil filtrate}}{\text{sample, g}}$$

3.5 DETERMINATION OF EXCHANGEABLE ACIDITY AND EXCHANGEABLE ALUMINUM USING 1 N POTASSIUM CHLORIDE

1. Principle of the Method

1.1. The acidity measured by the barium chloride (BaCl$_2$)-TEA method bears very little relationship to that to which plant roots react. The 1 N potassium chloride (KCl) method extracts the acidity exchangeable at the existing soil pH and consists primarily of aluminum (Al) and some hydrogen (H) (Sumner, 1992b). It is termed the "active" acidity in soil and determines to a substantial extent whether or not roots will grow in an acid soil.

2. Range and Sensitivity

2.1. This is a reliable and convenient method that is quite accurate. Exchangeable acidity as little as 0.05 meq per 100 g is readily determined.

3. Interferences

3.1. There are no interferences.

4. Apparatus

4.1. Beakers, 100-mL glass.
4.2. Buchner funnel (5.5 cm) and vacuum flask.
4.3. Whatman No 42 filter paper, 5.5 cm.
4.4. Buret, 50 mL.

5. Reagents (use reagent grade chemicals and pure water)

5.1. *Replacing Solution* (1 N potassium chloride): Weigh 74.56 g potassium chloride (KCl) in 1,000 mL water.
5.2. *Aluminum Complexing Solution* (1 N potassium fluoride): Titrate 58.1 g potassium fluoride (KF) per 1,000 mL to a phenolphathalein (see 5.5) endpoint with 0.1N sodium hydroxide (NaOH) (see 5.4).
5.3. *Hydrochloric Acid (HCl):* Approximately 0.1 N standardised.
5.4. *Sodium Hydroxide (NaOH):* Approximately 0.1 N standardised.
5.5. *Phenolphthalein Solution:* Weigh 1 g phenolphthalein in 100 mL ethanol.

6. Procedure

6.1. Scoop 10 cm^3 air dry <10-mesh soil into 100-mL beaker, add 25 mL 1 N KCl (see 5.1), mix, and allow to stand for 30 minutes. Transfer mixture to Buchner filtration system and add 5 × 25 mL aliquots of 1 N KCl (see 5.1) to give a total volume of 150 mL.
6.2. Titrate filtrate, after adding four to five drops of phenolphthalein (see 5.5) with 0.1 N NaOH (see 5.4) to the first permanent pink endpoint. This titre gives exchangeable acidity.
6.3. Add 10 mL 1 N KF (see 5.2) and titrate with 0.1 N HCl (see 5.3) until pink color disappears. Wait 30 minutes and add additional HCl to a clear endpoint. This titre gives exchangeable Al.

7. Calculations

$$\text{meq KCl acidity} = \frac{(\text{mL NaOH sample} - \text{mL NaOH blank}) \times N \times 100}{\text{sample, g}}$$

$$\text{meq KCl exchangeable Al} = \frac{\text{mL HCl} \times N \times 100}{\text{sample, g}}$$

meq H = KCl exchangeable acidity − KCl exchangeable Al

3.6 REFERENCES

Alley, M. M. and L. W. Zelazny. 1987. Soil acidity: Soil pH and lime needs, In: *Soil Testing: Sampling, Correlation, Calibration, and Interpretation,* SSSA Special Publication 21. J. R. Brown (ed.). Madison, WI: Soil Science Society of America, 65–72.

Anon. 1994. pH: 1:2 soil to water ratio, In: *Plant, Soils, and Water Reference Methods for the Western Region,* Western Regional Extension Publication WREP 125, R. G. Gavlak, D. A. Horneck, and R. O. Miller, (eds.). Fairbanks. AK: University of Alaska, 17–18.

Anon. 1996. Soil reaction and liming. Chapter 2. *Soil Fertility Manual.* Norcross, GA: Potash and Phosphate Institute, 2:1–9.

Anon. 1998. *Soil Test Levels in North America: Summary Update.* PPI/PPIC/FAR Technical Bulletin 1998-3. Norcross, GA: Potash & Phosphate Institute.

Black, C. A. 1993. *Soil Fertility Evaluation and Control,* Boca Raton, FL: Lewis Publishers, 647–728.

Coleman, N. T. and G. W. Thomas. 1967. The basic chemistry of soil acidity, In: *Soil Acidity and Liming* R. W. Pearson and F. Adams (eds.). Madison, WI: American Society of Agronomy, 1–41.

Conyers, M. K. and B. G. Darey. 1988. Observations in some routine methods for soil pH determination. *Soil Sci.* 145:29–36.

Fotyma, M., T. Jadczyszyn, and G. Jozefaciuk. 1998. Hundredth molar calcium chloride extraction procedure, Part II: Calibration with conventional soil testing methods for pH. *Commun. Soil Sci. Plant Anal.* 29:1625–1632.

Hendershot, W. H., H. Lalande, and M. Duquette. 1993. Soil reaction and exchangeable acidity, In: *Soil Sampling and Methods of Analysis,* M. R. Carter (ed.). Boca Raton, FL: Lewis (CRC) Press, 141–145.

Houba, V. J. G. and I. Novozamsky. 1998. Influence of storage time and temperature of air-dried soils on pH and extractable nutrients using 0.01 M $CaCl_2$. *Fresenius J. Anal. Chem.* 360:362–365.

Kalra, Y. P. 1995. Determination of pH of soils by different methods: Collaborative study. *J. Assoc. Off. Anal. Chem.* 78:310–324.

Kalra, Y. P. 1996. Soil pH: First soil analysis methods validated by the AOAC International. *J. For. Res.* 1:61–64.

Maynard, D. N. and G. J. Hochmuth. 1997. *Knott's Handbook for Vegetable Growers.* 4th ed. New York: John Wiley & Sons.

Mehlich, A. 1939. Use of triethanalamine acetate-barium hydroxide buffer for the determination of some base exchange properties and lime requirement of soil. *Soil Sci. Amer. Proc.* 3:162–166.

Mehlich, A. 1953. Rapid determination of cation and anion exchange properties and pH of soils. *J. Assoc. Off. Agr. Chem.* 36:445–457.

Meier, P., A. Lohrum, and J. Gareiss. 1989. *Practice and Theory of pH Measurement: An Outline of pH Measurement, Information, and Practical Hints.* Urdorf, Switzerland: Ingold Messtechnik AG.

McLean, E. O. 1973. Testing soils for pH and lime requirement, In: *Soil Testing and Plant Analysis,* (revised ed.), L. M. Walsh and J. D. Beaton (eds.). Madison, WI: Soil Science Society of America, 78–95.

Pearson, R. W. and F. Adams (eds.). 1967. *Soil Acidity and Liming.* Madison, WI: American Society of Agronomy.

Peech, M. 1965. Hydrogen-ion activity, In: *Methods of Soil Analysis,* Part 2. Agronomy No. 9. C. A. Black (ed.). Madison, WI: American Society of Agronomy, 914–926.

Plank, C. O. 1992a. Determination of soil pH on 0.01 M $CaCl_2$, In: *Reference Soil and Media Diagnostic Procedures for the Southern Region of the U.S.,* Southern Cooperative Series Bulletin No. 374, S. J. Donohue (ed.). Blacksburg, VA: Virginia Agricultural Experiment Station, 1–3.

Plank, C. O. 1992b. Determination of soil pH in 1 N KCl, In: *Reference Soil and Media Diagnostic Procedures for the Southern Region of the U.S.*, Southern Cooperative Series Bulletin No. 374. S. J. Donohue (ed.). Blacksburg, VA: Virginia Agricultural Experiment Station.

Reid, K. (ed.). 1998. Soil pH, liming and acidification, In: *Soil Fertility Handbook*. Toronto: Ministry of Agriculture, Food, and Rural Affairs, 7–82.

Sartoretto, P. 1991. *The pH factor.* Dayton, NJ: W-A-Clearly Chemical.

Schofield, R. K. and A. W. Taylor. 1955. The measurement of soil pH. *Soil Sci. Soc. Amer. Proc.* 19:164–167.

Sims, J. T. and D. Eckert. 1995. Recommended soil pH and lime requirement tests. In: *Recommended Soil Testing Procedures for the Northeastern U.S.*, 2nd ed. Northeast Regional Publication. J. T. Sims and Ann Wolf (eds.). Agricultural Experiment Station Bulletin 493. Newark, DE: University of Delaware, 35–40.

Slattery, W. J., M. K. Conyers, and R. L. Aitken. 1999. Soil pH, aluminium, manganese, and lime requirement. In: *Soil Analysis: An Interpretation Manual*, K. I. Peverill, L. A. Sparrow, and D. J. Reuter (eds.). Collingwood, Australia: CSIRO Publishing, 103–128.

Summer, M. E. 1992a. Determination of exchangeable acidity using $BaCl_2$-TEA buffer, in: *Reference Soil and Media Diagnostic Procedures for the Southern Region of the U.S.*, Southern Cooperative Series Bulletin No. 374, S. J. Donohue (ed.). Blacksburg, VA: Viriginia Agricultural Experiment Station, VPI & SU, 39–40.

Summer, M. E. 1992b. Determination of exchangeable acidity and exchangeable Al using 1N KCl. In: *Reference Soil and Media Diagnostic Procedures for the Southern Region of the U.S.*, Southern Cooperative Series Bulletin No. 374, S. J. Donohue (ed.). Blacksburg, VA: Virginia Agricultural Experiment Station, VPI & SU, 41–42.

Summer, M. E. 1994. Measurement of soil pH: Problems and solutions. *Commun. Soil Sci. Plant Anal.* 25:859–879.

Thomas, G. W. 1996. Soil pH and soil acidity, In: *Methods of Soil Analysis, Part 3, Chemical Methods*, SSSA Book Series No. 5. D. L. Sparks (ed.). Madison, WI: Soil Science Society of America, 475–490.

van Lierop, W. 1990. Soil pH and lime requirement determinations, In: *Soil Testing and Plant Analysis*. SSSA Book Series No. 3, R. L. Westerman (ed.). Madison, WI: Soil Science Society of America, 73–126.

Watson, M. E. and J. R. Brown. 1998. pH and lime requirement, In: *Recommended Chemical Soil Test Procedures for the North Central Region*, North Central Regional Research Publication No. 22 (revised), J. R. Brown (ed.). Columbia, MO: Missouri Agricultural Experiment Station SB 1001, University of Missouri, 13–16.

Woodruff, C. M. 1961. Bromocresol purple as an indicator of soil pH. *Soil Sci.* 91:272.

Woodruff, C. M. 1967. Crop response to lime in the midwestern U.S., In: *Soil Acidity and Liming*. R. W. Pearson and F. Adams (eds.). Madison, WI: American Society of Agronomy, 207–227.

4 Buffer pH and Lime Requirement

The soil pH is buffered by the solid phase components in the soil, the clay minerals, hydroxy-aluminum monomers and polymers, and organic matter, and in alkaline soils, undissoved carbonate compounds (Schofield and Taylor, 1955; Schwertmann and Jackson, 1964; Coleman and Thomas, 1967; Alley and Zelazny, 1987; Thomas, 1982, 1996). A change is soil pH is the result of a shift in the equilibrium when an acid, such as an ammonium (NH_4)-nitrogen (N) fertilizer, or a base, such as agricultural limestone, are added to the soil. The degree of change is not predictable on the basis of a simple acid-base calulation which considers only the acid or base in the soil solution.

Lime requirement (LR) test procedures applied to an acid soil are designed to measure the reserve acidity so that sufficient liming material is added to adjust the soil pH to a particular level (Keeney and Corey, 1963; van Lierop, 1990; Anon., 1994; Sims and Eckert, 1995; Sims, 1996; Reid, 1998; Slattery et al., 1999). Factors that are taken into consideration are the desired change in pH based on the volume of soil (normally described by depth) to be treated, and the characteristics of the liming material (fineness and total neutralizing power [TNP]) (see AGLIME Facts, PPI Reference No. 96172).[1] Two of the LR test procedures given in this chapter, the Adams-Evans and SMP buffer methods, are to be used for different soil characteristics, course-textured soils for the former and heavier-textured soils for the latter. If either test is used for soils other than so designated, over and under estimates of the LR will be made.

Comparisons among LR methods have been studied by McLean et al. (1966), Yuan (1975), Fox (1980), and Doerge and Gardner (1988).

Lime requirement concepts are discussed by Sea Tran and van Lierop (1993), Black (1993), and Mikkelsen and Camberato (1995) and is discussed in the PPI *Soil Fertility Manual* (Anon., 1996).

4.1 ADAMS-EVANS LIME BUFFER

1. Principle of the Method

1.1. This procedure describes the determination of the lime requirement (LR) of a soil by the Adams-Evans buffer method (Adams and Evans, 1962). The method was developed for non-montmorillonitic, low organic matter soils where the amounts of limestone needed are small and the possibility of damage from overliming exists. The LR of an acid soil is defined by this procedure as the amount of limestone or other base required to change an acid condition to a less acid condition (a maximum pH_W of 6.5).

---

[1] Potash & Phosphate Institute, 655 Engineering Drive, Suite 110, Norcross, GA 30092-2837

1.2. The Adams-Evans LR method is based on separate measures of soil pH determined in water and buffer pH determined in the Adams-Evans buffer. Soil pH is used as a measure of acid saturation of the soil, designated "H-sat$_1$" below (Hajek et al. 1972), according to the following:

$$\text{Measured soil pH} = 7.79 - 5.55 \, (\text{H-sat}_1) + 2.27 \, (\text{H-sat}_1)^2$$

where H saturation is expressed as a fraction of cation exchange capacity (CEC). Buffer pH is used as a measure of soil acids, designated "soil H" below (Adams and Evans, 1972; Hajek et al., 1972), according to the equation:

$$\text{Soil H} = 8 \, (8.00 - \text{buffer pH})$$

for a 10 g soil sample in 10 mL water plus 10 mL buffer where "soil H" is in meq 100 g^{-1} of soil. A pH change of 0.01 in 20 mL of solution (10 mL water plus 10 mL buffer) is caused by 0.008 meq of acid at a pH level between 7.0 and 8.0. Cation exchange capacity is calculated by using "H-sat$_1$" and "soil H" according to the equation:

$$\text{CEC} = \text{Soil H/H-sat}_1.$$

The desired soil pH (not to exceed 6.5) is expressed in terms of acid saturation (designated "H-sat$_2$" below), according to the following:

$$\text{Desired soil pH} = 7.79 - 5.55 \, (\text{H-sat}_2) + 2.27 \, (\text{H-sat}_1)^2.$$

2. Range and Sensitivity

2.1. The Adams-Evans buffer method is very reliable for soils with relatively small amounts of exchangeable acidity (max. = 8 meq 100 g^{-1}). The procedure provides a fairly high degree of accuracy for estimating LRs to reach pH 6.5 or less.

2.2. Sensitivity for the lime requirement determination is within 500 limestone lb A^{-1}.

3. Interferences

3.1. No significant interferences.

4. Sensitivity

4.1. A sensitivity of 0.01 in pH units of the buffer-soil slurry is needed.

5. Apparatus

5.1. No. 10 (2-mm opening) sieve.
5.2. 10-cm^3 scoop, volumetric.
5.3. 50-mL cup, glass, plastic, or waxed paper.
5.4. Pipette, 10-mL capacity.
5.5. Mechanical shaker (180 oscillations per minute) or stirrer.

Buffer pH and Lime Requirement

5.6. pH meter with reproducibility to at least 0.01 pH unit and a glass electrode paired with a calomel reference electrode.
5.7. Glassware and dispensing apparatus for the preparation and dispensing of Adams-Evans buffer.
5.8. Analytical balance.

6. Reagents (use reagent grade chemicals and pure water)

6.1. *pH Buffer pH 4.0:* Commercially prepared pH buffer solution, or weigh 11.8060 g citric acid ($C_6H_8O_7$) and 10.9468 g disodium phosphate ($Na_2HPO_4 \cdot 12H_2O$) into a 1,000-mL volumetric flask and bring to volume with water.
6.2. *pH Buffer pH 7.0:* Commercially prepared pH buffer solution, or weigh 3.3910 g citric acid ($C_6H_8O_7$) and 23.3844 g disodium phosphate ($Na_2HPO_4 \cdot 12H_2O$) into a 1,000-mL volumetric flask and bring to volume with water.
6.3. *Adams-Evans Lime Buffer Solution:* Weigh 74 g potassium chloride (KCl) into a 1,000-mL volumetric flask containing 500 mL water, add 10.5 g potassium hydroxide (KOH), and stir to bring into solution. Add 20 g p-nitrophenol ($HO \cdot C_6H_4 \cdot NO_2$) and continue to stir. Add 15 g boric acid (H_3BO_3). Stir and heat, if necessary, to bring into solution. Bring to volume with water when cool.

7. Procedure

7.1. *Determination:* Scoop 10 cm^3 air-dry <10-mesh (2-mm) soil into a 50-mL cup. Add 10 mL water and mix for five seconds. Wait for 10 minutes and read the soil pH while stirring using a calibrated pH meter (see 8.1). Only on samples with pH$_W$ less than 6.4, add 10 mL Adams-Evans Buffer Solution (see 6.3) to the cup. Shake 10 minutes (see 5.5) or stir intermittently for 10 minutes. Let stand for 30 minutes. Read the soil-buffer pH on a standardized pH meter (see 8.2). Stir the soil suspension just prior to reading the pH. Read the pH to the nearest 0.01 pH unit.

8. Calibration

8.1. For calibration of the pH meter when determining water pH, follow the manufacturer's instructions using the pH buffers 4.0 and 7.0 (see 6.1 and 6.2).
8.2. The pH meter is adjusted to read pH 8.00 in an equal volume Adams-Evans Buffer (see 6.3) and water.

9. Calculation

9.1. The Adams-Evans buffer method assumes that agricultural-grade limestone is about 2/3 effective in neutralizing acidity up to a soil pH of about 6.5 and allows for this by using a correction factor of 1.5. Thus, the LR is the product of the following equation (see 1.2):

$$(\text{Soil H})/\text{H-sat}_1 \times (\text{H-sat}_1 - \text{H-sat}_2) \times 1.5$$

or for 10 g soil in 10 mL water plus 10 mL buffer (see 6.3), it is $CaCO_3$ (tons A^{-1}) = $8[(8.00-\text{buffer pH})/\text{H-sat}_1] \times (\text{H-sat}_1 - \text{H-sat}_2) \times 1.5$.

10. Effects of Storage

10.1. Air-dry soils may be stored several months in closed containers without affecting the pH$_{Adams}$ measurement.

10.2. If the pH meter and electrodes are not to be used for extended periods of time, the storage instructions published by the instrument manufacturer should be followed.

11. Interpretation

11.1. The Adams-Evans buffer method was developed for soils that have a maximum soil hydrogen (H) content of 8.00 meq 100 g^{-1}, and which have H-sat$_1$ of 1.00 at about pH 4.5. However, it can be used with soils with more H by adding less than 10 g soil to 10 mL water plus 10 mL buffer and multiplying by the appropriate dilution factor. It also can be used with soils that have pH values below 4.5 when H-sat$_1$ is 1.00 by changing the intercept of the pH equation by the appropriate amount (see 1.2). For example, a soil that has a pH of 4.0 when H-sat is 1.00 has the following relationship between pH and H-saturation:

$$\text{Soil pH} = 7.29 - 5.55 \, (\text{H-sat}) + 2.27 \, (\text{H-sat})^2.$$

11.2. The LR for low CEC soils (and with a pH of about 4.5 when H-saturated) can be determined in the following Lime Requirement Table. This table is based on the Adams-Evans buffer values:

Limestone (Ag-ground, TNP = in 1,000 lbs A^{-1}) to raise soil pH to 6.5 to a soil depth of 6-2/3 Inches

Buffer pH	\multicolumn{10}{c}{Soil Water pH}
	6.2	6.0	5.8	5.6	5.4	5.2	5.0	4.8	4.6	4.4
	\multicolumn{10}{c}{1,000 1b A^{-1}}
7.95	0	0	0	0	2	2	2	2	2	2
7.90	0	0	0	0	2	2	2	2	2	2
7.85	0	0	2	2	2	2	2	2	2	2
7.80	0	1	2	2	2	2	2	2	2	2
7.75	0	1	2	2	2	2	2	2	2	2
7.70	1	1	2	2	2	2	2	2	2	
7.65	1	1	2	2	2	2	3	3	3	3
7.60	1	2	2	2	2	3	3	3	3	4
7.55	1	2	2	2	3	3	3	4	4	4
7.50	1	2	3	3	3	3	4	4	4	5
7.45	2	2	3	3	3	4	4	4	5	5
7.40	2	2	3	3	4	4	4	5	5	5
7.35	2	2	3	4	4	5	5	5	5	6
7.30	2	3	4	4	4	5	5	5	6	6
7.25	2	3	4	4	5	5	5	6	6	7

7.20	2	3	4	5	5	6	6	6	7	7
7.15	2	3	4	5	5	6	6	7	7	8
7.10	2	3	5	5	4	7	7	7	8	8
7.05	3	4	5	5	6	7	7	7	8	8
7.00	3	4	5	6	7	7	8	8	8	9

4.2 SMP LIME BUFFER: ORIGINAL AND DOUBLE-BUFFER ADAPTATION

1. Principle of the Method

1.1. This procedure describes the determination of the lime requirement (LR) of a soil by the Shoemaker, McLean, and Pratt (SMP) buffer method (Shoemaker et al., 1962, McLean et al., 1977, 1978; Anon, 1994; Sims and Eckert, 1995; Watson and Brown, 1998). The LR of an acid soil is defined as "the amount of limestone or other bases which when incorporated within a given depth of acid soil increases the pH to some selected level." It is expressed as the calcium carbonate ($CaCO_3$) equivalent in tons per acre (tons A^{-1}) of plowed soil to a depth of 8 inches (20 cm) which is equivalent to 2,240 kg ha^{-1} $CaCO_3$ to the same depth or to 1.67 meq 100^{-1} g of soil. An acre 8-inch furrow slice of soil is assumed to weigh 2,400,000 pounds.

1.2. The SMP buffer method measures the change in pH of a buffer caused by the acids in the soil, and this change in buffer pH is a measure of the LR of the soil. The double-buffer adaptation involves the individual slope of the buffer-indicated versus actual LR curve for a given soil instead of a mean slope involved in the original method (Shoemaker et al., 1962; McLean et al., 1977).

2. Range and Sensitivity

2.1. The SMP buffer method is very reliable for soils with a greater than 4,480 kg ha^{-1} (2 tons A^{-1}) LR. It is also well adapted for acid soils with a pH below 5.8 containing less than 10% organic matter and having appreciable quantities of soluble aluminum (Al) (McLean et al., 1966).

2.2. The original method was never considered to be very accurate for soils with LR less than 4,480 kg ha^{-1} (2 tons A^{-1}) because of random variation of buffer-indicated versus actual LR in this range (Shoemaker et al., 1962). Also, on mineral soils of high organic matter and high levels of extractable Al the original SMP method indicates LRs which are lower than the actual amounts required. However, an adaptation of the original method for use on organic soils was included some time ago (van Lierop, 1983). More recently the double-buffer adaptation originally suggested by Yuan (1974) has been developed to improve some of the shortcomings of the original method (McLean et al., 1977, 1978).

3. Interferences

3.1. Increased time of soil contact with the buffer results in lower buffer pH, and therefore, a greater LR.

3.2. Organic matter and LR are highly correlated (Kenney and Corey, 1963), because when organic matter increases, more acidic cations accumulate on the exchange sites

When organic matter is very high, especially when Al is complexed with it, a portion of the LR may not be measured by the original SMP method (McLean et al., 1978).

4. Sensitivity

4.1. A sensitivity of 0.1 pH unit is needed for the interpretation of this analysis by the original procedure (Shoemaker et al., 1962). But the double-buffer adaptation calls for pH readings to the nearest 0.01 pH unit (McLean et al., 1977). A difference of 0.1 pH unit in the original method results in a difference of 0.4 to 0.6 tons lime A^{-1} for organic soils limed to pH 5.2 and 0.5 to 0.9 tons A^{-1} for mineral soils limed to near neutral pH. Similarly, a difference of 0.1 pH unit in one of the two buffers in the double-buffer adaptation may result in a difference in lime requirement of less than 0.1 ton A^{-1} for mineral soils of low LR to more than 0.5 ton A^{-1} for soils of high LR.

4.2. The SMP buffer method is not applicable to soils which have a low buffer capacity. By adding the SMP buffer solution (pH 7.5) to the soil-water suspension used to determine the pH water, the pH will increase to a level which can not be used in combination with the table mentioned under Section 12.3.

5. Apparatus

5.1. No. 10 (2-mm opening) sieve.
5.2. Scoop, 4.25-cm³ volumetric.
5.3. Cup, 50-mL (glass, plastic, or waxed paper of similar size).
5.4. Pipette, 5-mL capacity.
5.5. Mechanical shaker.
5.6. pH meter with reproducibility to 0.01 pH unit and glass electrode paired with a calomel reference electrode.
5.7. Glassware and dispensing apparatus for the preparation and dispensing SMP buffer.
5.8. Analytical balance.

6. Reagents (use reagent grade chemicals and pure water)

6.1. *pH Buffer pH 4.0:* Commercially prepared pH buffer solution, or weigh 11.8060 g citric acid ($C_6H_8O_7$) and 10.9468 g disodium phosphate ($Na_2HPO_4·12H_2O$) into a 1,000 mL volumetric flask and bring to volume with water.

6.2. *pH Buffer pH 7.0:* Commercially prepared pH buffer solution, or weigh 3.3910 g citric acid ($C_6H_8O_7$) and 23.3844 g disodium phosphate ($Na_2HPO_4·12H_2O$) into a 1,000-mL volumetric flask and bring to volume with water.

6.3. *SMP Buffer Solution:* Weigh into an 18-L bottle 32.4 g paranitrophenol, 54.0 g potassium chromate ($KCrO_4$), and 955.8 g calcium chloride dihydrate. ($CaCl_2·2H_2O$). Add approximately 9 L water. Shake vigorously as water is added and continue shaking for a few minutes to prevent formation of a crust over the salts. Weigh 36.0 g calcium acetate [$Ca(C_2H_3O_2)_2·H_2O$] into a separate container and dissolve in approximately 5 L water. Add latter solution to the former, shaking as they are combined. Shake every 15 or 20 minutes for two or three hours. Add 45 mL triethanolamine; again shaking as the addition is made. Shake periodically until completely dissolved. This takes approximately eight hours. Dilute to 18 L with distilled

Buffer pH and Lime Requirement

water. Adjust to pH 7.5 with 15% sodium hydroxide (NaOH) using a standardized pH meter. Filter through a fiber glass sheet or cotton mat. Connect an air inlet with 1-inch by 12-inch cylinder of Drierite®, 1- by 12-inch cylinder of Ascarite®, and 1-by 12-inch cylinder of Drierite in series to protect against contamination with carbon dioxide (CO_2) and water vapor. Although less tedious procedures may be used for preparing small quantities of the buffer solution, the above procedure has been found to be most satisfactory for preparing bulk quantities of the buffer solution.

7. Procedure

7.1. *Determination:* Weigh 5 g or scoop 4.25 cm^3 of air-dry <10-mesh (2-mm) soil into a 50-mL cup in a tray designed for a mechanical shaker. Add 5 mL water, shake or stir one minute, let stand 10 minutes, and read pH (see 9.1) in water with slight swirling of the slurry. Add 10 mL SMP buffer adjusted to pH 7.5 (see 6.3) to the above soil suspension, shake on a mechanical shaker at >180 oscillations per minute for 10 minutes, open the lid of the shaker, and let stand 30 minutes. Read buffer pH (pH_1) on carefully calibrated pH meter to nearest 0.01 pH unit (see 9.1). A 15 minute shaking time and 15 minute standing time may be used if more adaptable to the soil testing routine, since this gives essentially the same results as the 10 minute shaking plus 30 minute standing times.

7.2. If the original SMP (one-buffer) method is to be used, select the LR from the table (see 12.3) based on the buffer pH obtained.

7.3. If the double-buffer adaptation is to be used, continue the procedure as follows: Using an automatic pipette, add to the above soil-buffer suspension an aliquot of hydrochloric acid (HCl) equivalent to the amount required to decrease a 10 mL aliquot of pH 7.5 buffer to pH 6.0 (1 mL 0.206M HCl–0.206 meq). Repeat the 10 minute shaking, 30 minute standing (or 15 minutes of shaking plus 15 minutes standing), and reading of soil-buffer pH (pH_2). Use the double-buffer formula and mathematical function indicated is 8.1 to convert pH readings of the quick test method to actual LR values. The procedure for use of the double-buffer adaptation has not yet been worked out for liming organic soils to pH 5.2.

8. Computations

8.1. d in meq per 5 g soil =

$$\Delta pH_2 \times \frac{\Delta d_2^o}{\Delta pH_2^o} + \left[\left[\Delta pH_1 \times \frac{\Delta d_1^o}{\Delta pH_1^o} - \Delta pH_2 \times \frac{\Delta d_2^o}{\Delta pH_2^o} \right] \times \left[\frac{6.5 - pH_2}{pH_1 - pH_2} \right] \right]$$

$$(d_2) \qquad\qquad (d_1) \qquad\qquad (d_2) \qquad\qquad (\beta)$$

where:
1) pH_1 is soil-buffer pH in pH 7.5 buffer.
2) pH_2 is soil-buffer pH in pH 6.0 buffer.
3) ΔpH_1 = 7.5 - pH_1.
4) ΔpH_2 = 6.0 - pH_2.
5) Δd_1^o = change in acidity per unit change in pH of 10 mL ΔpH_1^o of pH 7.5 buffer by titration ~0.137 meq per unit pH.

6) Δd_2^o = change in acidity per unit change in pH of 10 mL ΔpH_2^o of pH 6.0 buffer by titration ~0.129 meq per unit pH.

7) 6.5 = pH to which soil is to be limed (any pH may be chosen).

8.2. LR in meq 100 g^{-1} soil = 1.69y − 0.86

where: y = 20d and d is the acidity in meq 5 g^{-1} soil measured by the double-buffer (or one-buffer two-pH) procedure.

The equation derived from the regression of buffer-indicated versus actual [Ca(OH)$_2$ titrated] LR corrects for less than complete reaction with the soil acidity in 10 minutes shaking and 30 minutes standing time (or 15 minutes shaking plus 15 minutes standing). If a soil is so acid that 5 g depresses the pH 6.0 buffer below pH 4.8, 4 g can be used with a multiple of 25 instead of 20.

9. Calibration and Standards

9.1. The pH meter is calibrated using prepared (see 6.1 and 6.2) or commercially prepared buffer solutions of pH 4.0 and pH 7.0 according to the instrument instruction manual.

10. Calculations

10.1. Lime requirements computed from double-buffer, quick-test formulas and expressed as meq 100 g^{-1} of soil are converted to tons CaCO$_3$ A^{-1} 8-inch depth of soil (2,400,000 lbs) by multiplying LR by 0.6, or per acre 6-2/3 inch (2,000,000 lbs) by multiplying by 0.5.

11. Effects of Storage

11.1. Air-dry soils may be stored several months in closed containers without appreciable effect on the pH$_{SMP}$ measurement.

11.2. If the pH meter and electrodes are not to be used for extended periods of time, the instructions for storage published by the instrument manufacturer should be followed.

12. Interpretation

12.1. The regular (single-buffer) SMP method is probably the most satisfactory compromise between simplicity of measurement and reasonable accuracy for soils of a wide range in lime requirement (McLean et al., 1978). As indicated above, the LR for any soil can be determined using the table in 12.3 based on the SMP soil-buffer pH values which gives the LR in terms of tons A^{-1} of agricultural ground limestone of total neutralizing power (TNP) or CaCO$_3$ equivalent of 90% or above and an 8-inch plow depth (2,400,000 lbs) to increase soil pH to selected levels.

12.2. The double-buffer adaptation is somewhat more accurate for all soils, but is especially so for soils of relatively low LR (McLean et al., 1977) and probably so for acid mineral soils of relatively high organic matter content.

12.3. Amounts of limestone required to bring mineral and organic soils to the indicated pH according to soil-buffer pH (tons A^{-1} 8-inch depth of soil) are presented in the following table:

Soil-Buffer pH	Mineral Soils				Organic Soils
	7.0 Pure CaCO$_3$	7.0	6.5	6.0	5.2
			Ag-ground Limestone* tons A^{-1}		
6.8	1.1	1.4	1.2	1.0	0.7
6.7	1.8	2.4	2.1	1.7	1.3
6.6	2.4	3.4	2.9	2.4	1.8
6.5	3.1	4.5	3.8	3.1	2.4
6.4	4.0	5.5	4.7	3.8	2.9
6.3	4.7	6.5	5.5	4.5	3.5
6.2	5.4	7.5	6.4	5.2	4.0
6.1	6.0	8.6	7.2	5.9	4.6
6.0	6.8	9.6	8.1	6.6	5.1
5.9	7.7	10.6	9.0	7.3	5.7
5.8	8.3	11.7	9.8	8.0	6.2
5.7	9.0	12.7	10.7	8.7	6.7
5.6	9.7	13.7	11.6	9.4	7.3
5.5	10.4	14.8	12.5	10.2	7.8
5.4	11.3	15.8	13.4	10.9	8.4
5.3	11.9	16.9	14.2	11.6	8.9
5.2	12.7	17.9	15.1	12.3	9.4
5.1	13.5	19.0	16.0	13.0	10.0
5.0	14.2	20.0	16.9	13.7	10.5
4.9	15.0	21.1	17.8	14.4	11.0
4.8	15.6	22.1	18.6	15.1	11.6

*Ag-ground limestone of 90% plus total neutralizing power (TNP) or CaCO$_3$ equivalent, and fineness of 40% < 100 mesh, 50% < 60 mesh, 70% < 20 mesh, and 95% < 8 mesh.

4.3 EXCHANGEABLE ACIDITY AND LIME REQUIREMENT BY THE MEHLICH BUFFER-pH METHOD

1. Principle of the Method

1.1. This procedure describes the determination of weight per volume, soil pH, and buffer pH, including the calculation of soil acidity (AC), estimation of unbuffered salt exchangeable acidity (ACe) and lime requirement (LR) (Mehlich, 1939, 1953, 1976). Methods now in use are calibrated mainly against soil pH (Shoemaker et al., 1962), against percentage of base unsaturation (Adams and Evans, 1962) and use of a double buffer for sandy soils (Yuan, 1974, 1975). However, in view of the importance of unbuffered salt exchangeable acidity in relation to liming (Kamprath, 1970; Evans and Kamprath, 1970; Mehlich et al., 1976), there is a need for a buffer pH method primarily calibrated against ACe and with special reference to exchangeable aluminum (Al^{3+}). In addition, the buffer pH acidity (AC_b) was standardized against crop response to liming under greenhouse and field conditions (Mehlich, 1976). For the estimation of ACe, determined AC was used in regression equations for mineral soils and histosols. The main objective of the procedure for LR was to determine the quantity of lime needed to neutralize a portion or all the ACe required for optimum plant growth. This quantity was expressed in a curvilinear function of AC for mineral soils, including those having histic epipedon and for histosols as a function of AC used in the regression equation for calculating ACe.

1.2. Measurements of weight per volume in conjunction with the percentage of organic matter provide the indexes for differentiating between mineral soils and histosols for calculating LR. The determination of soil pH provides the index to the need for liming depending on acid (Al^{3+}) tolerance of crops and major soil differences.

2. Range and Sensitivity

2.1. The capacity of the buffer (pH 6.6-4.0) is equivalent to 10 meq $CaCO_3$ 100 cm^{-3}, 20 metric tons (MT) limestone ha^{-1}, or 20,000 lb A^{-1}. Provisions in the procedure allow this capacity to be doubled.

2.2. Sensitivity per 0.1 pH depression of buffer is 0.4 meq $CaCO_3$ 100 cm^{-3}, 0.4 MT ha^{-1}, or 400 lb A^{-1}.

3. Interferences

3.1. Buffer pH of soil suspension should be read after 60 minutes standing. All measurements should be made within the same day.

4. Sensitivity

4.1. A sensitivity of 0.1 pH unit is adequate for soils having a LR greater than 2 MT ha^{-1}, while for a lower LR, a sensitivity of 0.05 pH units would be desirable.

5. Apparatus

5.1. No. 10, 2-mm ISO standard sieve.
5.2. 10-cm^3 scoop, volumetric.
5.3. 50-mL cup, glass, plastic, or waxed paper cup.
5.4. Pipette, 10-mL capacity.
5.5. Mechanical shaker or stirrer (optional).
5.6. pH meter with reproducibility to 0.05 pH unit and a glass electrode paired with a calomel reference electrode.
5.7. Glassware and dispensing apparatus for the preparation and dispensing of 10 mL water and Mehlich buffer reagent.
5.8. Analytical balance.

6. Reagents (use reagent grade chemicals and pure water)

6.1. Sodium Glycerophosphate [(National Formulary) (N.F.) M.W. 315.15)]: The N. F. quality of sodium glycerophosphate [$Na_2C_3H_5(OH)_2PO_4 \cdot 5\ 1/2H_2O$] is satisfactory and considerably more economical than the crystal beta form (Source: Roussel Corporation, 155 E. 44th St., New York, NY 10017).

6.2. Buffer Solution: To about 1,500 mL water in a 2,000-mL volumetric flask or a 2,000-mL calibrated bottle, add 5 mL glacial acetic acid ($HC_2H_3O_2$) and 9 mL triethanolamine, or for ease of delivery, add 18 mL of an 1:1 aqueous mixture. Add 86 g ammonium chloride (NH_4Cl) and 40 g barium chloride ($BaCl_2 \cdot 2H_2O$) and dissolve. Dissolve separately 36 g sodium glycerophosphate in 400 mL water and transfer to the above 2,000-mL volumetric flask or bottle. Allow the endothermic reacted solution to reach room temperature and make up to volume with water and mix. Dilute an aliquot of the Buffer Solution with an equal volume of water and determine the pH. The pH of the Buffer Reagent should be 6.6. However, if it is above pH 6.64, add dropwise glacial $HC_2H_3O_2$. If it is below pH 6.56, add dropwise 1:1 aqueous triethanolamine. Check the concentration of the buffer by adding 10 mL 0.1 M HCl-$AlCl_3$ mixture [dissolve 4.024 g aluminum chloride ($AlCl_3 \cdot 6H_2O$) in 0.05 M hydrochloric acid (HCl)] to 10 mL buffer + 10 mL water and determine the pH. The correct pH obtained should be 4.1 ± 0.05.

6.3. pH Buffer pH 4.0: Commercially prepared pH buffer solution, or weigh 11.8060 g citric acid ($C_6H_8O_7$) and 10.9468 g disodium phosphate ($Na_2HPO_4 \cdot 12H_2O$) into a 1,000-mL volumetric flask and bring to volume with water.

6.4. pH Buffer pH 7.0: Commercially prepared pH buffer solution, or weigh 3.3910 g citric acid ($C_6H_8O_7$) and 23.3844 g disodium phosphate ($Na_2HPO_4 \cdot 12H_2O$) into a 1,000-mL volumetric flask and bring to volume with water.

7. Procedure

7.1. Water pH Measurement: Scoop 10 cm^3 of air-dry soil <10-mesh (2-mm) into a 50-mL cup (see 5.3). To obtain weight per volume, weigh the measured 10 cm^3 soil to the nearest 0.1 g, divide by 10, and express the results in g cm^{-3}. Add 10 mL water with sufficient force to mix with soil. After stirring for about 30 minutes, read soil pH while stirring (for poorly wettable histosols, add eight to 10 drops of ethanol).

7.3. Buffer pH Measurement: Add to the soil from the pH determination, 10 mL Buffer Solution (see 6.2) with sufficient force to mix. Read the buffer pH to the nearest 0.05 unit after 60 minutes while stirring. If it is desired to extend buffer capacity below pH 4.0, add an additional 10 mL Buffer Solution (see 6.2), equilibrate with stirring, and measure pH$_B$.

8. Calibration and Standards

8.1. Before measuring the buffer pH of the soil suspension, calibrate the pH meter to pH 6.6 in a mixture of 10 mL Buffer Solution (see 6.2), and 10 mL water.

9. Calculations for Exchangeable Acidity

9.1. Convert buffer pH (BpH) into buffer pH acidity (AC) as follows:

$$AC \text{ (meq 100 cm}^{-3}\text{) soil} = (6.6 - BpH)/0.25 \qquad [1]$$

(If a second 10 mL portion of buffer was used, multiply AC by 2.)

9.2. For unbuffered salt exchangeable acidity (ACe) based on AC of mineral soils, calculate:

$$ACe \text{ (meq 100 cm}^{-3}\text{)} = 0.54 + 0.96 \text{ (AC)} \qquad [2]$$

9.3. For ACe determination of histosols and mineral soils having histic epipedon, calculate:

$$ACe \text{ (meq 100 cm}^{-3}\text{)} = -7.4 + 1.6 \text{ (AC)} \qquad [3]$$

10. Calculations for Lime Requirement

10.1. Convert BpH into AC (see 9.1).

10.2. The lime requirement (LR) in the following equations may be expressed and is equivalent to meq $CaCO_3$ 100 cm^{-3} soil, metric tons (MT) ground limestone TNP = 90% ha^{-1} to a depth of 20 cm, or lbs A^{-1} (MT × 10^3).

10.3. *Mineral Soils:* for plants with slight to moderate tolerance for ACe and soil reaction in water of pH 5.8 to 6.5,

$$LR = 0.1 \text{ (AC)}^2 + AC \qquad [4]$$

10.4. *Mineral Soils:* for plants with low tolerance to ACe, and soil reaction in water <6.5, multiply results with Equation [3] by 1.5 or 2.0.

10.5. *Histosols or Mineral Soils with Histic Epipedon* (OM 20% and above): for soil reaction in water < pH 4.8 to 5.0 and 0.75 g W/V cm^{-3}, use Equation [2] × 1.3, viz.,

$$LR = [-7.4 + 1.6(AC)]1.3 \qquad [5]$$

10.6. *Mineral Soils High in Organic Matter* (OM 10 to 19%): for soil reaction in water <pH 5.3 to 5.5, and W/V within 0.75 to 0.95 g cm^{-3}, use Equation [4] with soils of sandy texture and Equation [3] with soils of silt and clay texture.

10.7. In all cases, when soil pH is below the indicated optimum, use 1 ton limestone ha^{-1} or its equivalent, even though AC is less than 0.5 meq 100 cm^{-3}.

11. Interpretation

11.1. The LR equations based on the proposed BpH method may be used in a computerized soil testing program. For manual use, the calculated LR values based on Equations [3] and [4] at 0.1 BpH intervals are recorded in the following table.

Buffer pH, AC and LR Conversion of Mineral and Organic Soils into MT ha^{-1} or lbs A^{-1} (MT × 10^3) of Ag. Ground Limestone with TNP = 90%.

Lime Requirement for Soils

BpH	AC	Mineral Equation [3]*	Organic Equation [4]*	BpH	AC	Mineral Equation [3]*	Organic Equation [4]*
6.6	0.0	0.0	0.0	5.2	5.6	8.7	2.0
6.5	0.4	0.4**	0.0	5.1	6.0	9.6	2.9
6.4	0.8	0.9	0.0	5.0	6.4	10.5	3.7
6.3	1.2	1.3	0.0	4.9	6.8	11.4	4.5
6.2	1.6	1.9	0.0	4.8	7.2	12.4	5.4
6.1	2.0	2.4	0.0	4.7	7.6	13.4	6.2
6.0	2.4	3.0	0.0	4.6	8.0	14.4	7.0
5.9	2.8	3.6	0.0	4.5	8.4	15.5	7.9
5.8	3.2	4.2	0.0	4.4	8.8	16.5	8.7
5.7	3.6	4.9	0.0	4.3	9.2	17.7	9.5
5.6	4.0	5.6	0.0	4.2	9.6	18.8	10.3
5.5	4.4	6.3	0.0	4.1	10.0	20.0	11.2
5.4	4.8	7.1	0.4**	4.0	10.4	21.2	12.0
5.3	5.2	7.9	1.2	3.9	10.8	22.5	12.8

*For Equations [3] and [4], see 10.3 to 10.6. For crops with high LR or very low tolerance to ACe, multiply the results of Equation [3] by a factor of 1.5 or 2.0.

**We suggest using 1 ton limestone ha^{-1} or 1,000 lbs A^{-1} when LR based on pH is indicated (see 10.3 to 10.6).

11.2. While liming needs are contingent on BpH, soil pH measured in a 1:1 soil:water ratio on a volume basis has been suggested as a criterion in the LR decision-making process. Soil pH levels measured in 1 N potassium chloride (KCl) and 0.01 M calcium chloride (CaCl$_2$·2H$_2$O) were found to deviate inconsistently from those measured in water. These deviations were largely related to the quantity and proportion of ACe to ACr, exchangeable Al^{3+} to H$^+$ and major soil components with respect to organic matter, layer silicates and sesquioxide hydrates. Schofield and Taylor (1955) introduced the use of 0.01 M CaCl$_2$·2H$_2$O in a 1:2 soil:salt solution ratio on a weight-to-volume basis as a measure of "lime potential." The authors determined pH in the supernatant liquid. Jackson (1958) stirred the soil suspension just before immersing the electrodes, and Peech (1965) placed the glass electrode into the partly settled suspension and the calomel electrode into the clear supernatant solution. In the case of acid ultisols, the relative decrease in pH from that obtained in a 1:1 soil:water suspension was, on the average, 1.0, 0.8, and 0.6 by the Jackson, Peech, and Schofield-Taylor procedures, respectively. With neutral to slightly acid soils, the total differences were in general less than one-half of the above. In view of the variability of soil pH obtained with varying salts due to procedural differences and soil properties, and because of the importance of maintaining uniformity of soil test results, measurement of pH in a 1:1 soil:water suspension by volume in conjunction with the proposed BpH method for LR is recommended.

4.4 REFERENCES

Adams, F. and C. E. Evans. 1962. A rapid method for measuring lime requirement of red-yellow podzolic soils. *Soil Sci. Soc. Amer. Proc.* 26:335–357.

Alley, M. M. and L. W. Zelazny. 1987. Soil acidity: Soil pH and lime needs, In: *Soil Testing: Sampling, Correlation, Calibration, and Interpretation,* SSSA Special Publication 21, J. R. Brown (ed.). Madison, WI: Soil Science Society of America.

Anon. 1994. Lime requirement: SMP buffer method, In: *Plant, Soils, and Water Reference Methods for the Western Region,* Western Regional Extension Publication WREP 125, R. G. Gavlak, D. A. Horneck, and R. O. Miller (eds.). Fairbanks, AK: University of Alaska.

Anon. 1996. Soil reaction and liming, In: *Soil Fertility Manual,* Chapter 2. Norcross, GA: Potash and Phosphate Institute, 2:1–9.

Black, C. A. 1993. *Soil Fertility Evaluation and Control.* Boca Raton, FL: Lewis Publishers.

Coleman, N. T. and G. W. Thomas. 1967. The basic chemistry of soil acidity, In: *Soil Acidity and Liming,* R. W. Pearson and F. Adams (eds.). Madison, WI: American Society of Agronomy, 1–41.

Doerge, T. A. and E. H. Gardner. 1988. Comparison of four methods for interpreting the Shoemaker-McLean-Pratt (SMP) lime requirement test. *Soil Sci. Soc. Amer. J.* 52:1054–1059.

Evans, C. E. and E. J. Kamprath. 1970. Lime response as related to percent Al saturation, solution Al and organic matter content. *Soil Sci. Soc. Amer. Proc.* 34:893–896.

Fox, R. H. 1980. Comparisons of several lime requirement methods for agricultural soils in Pennsylvania. *Commun. Soil Sci. Plant Anal.* 11:57–69.

Hajek, B. F., F. Adams, and J. T. Cope. 1972. Rapid determination of exchangeable bases, acidity and base saturation for soil characterization. *Soil Sci. Soc. Amer. Proc.* 36:436–438.

Jackson, M. L. 1958. *Soil Chemical Analysis.* Englewood Cliffs, NJ: Prentice-Hall.

Kamprath, E. J. 1970. Exchangeable aluminum as a criterion for liming leached mineral soils. *Soil Sci. Soc. Amer. Proc.* 34:252–254.

Keeney, D. R. and R. G. Corey. 1963. Factors affecting the lime requirements of Wisconsin soils. *Soil Sci. Soc. Amer. Proc.* 27:277–280.

McLean, E. O., S. W. Dumford, and F. Coronel. 1966. A comparison on several methods of determining lime requirements of soils. *Soil Sci. Soc. Amer. Proc.* 30:26–30.

McLean, E. O., J. F. Trierweiler, and D. J. Eckert. 1977. Improved SMP buffer method for determining lime requirement of acid soils. *Commun. Soil Sci. Plant Anal.* 8:667–675.

McLean, E. O., D. J. Eckert, G. Y. Reddy, and J. F. Trierweiler. 1978. Use of double-buffer and quicktest features for improving the SMP method for determining lime requirement of acid soils. *Soil Sci. Soc. Amer. J.* 42:311–316.

Mehlich, A. 1939. Use of triethanalamine acetate-barium hydroxide buffer for the determination of some base exchange properties and lime requirement of soil. *Soil Sci. Soc. Amer. Proc.* 3:162–166.

Mehlich, A. 1953. Rapid determination of cation and anion exchange properties and pHe of soils. *J. Assoc. Off. Agric. Chem.* 36:445–457.

Mehlich, A. 1976. New buffer pH method for rapid estimation of exchangeable acidity and lime requirement of soils. *Commun. Soil Sci. Plant Anal.* 7:637–652.

Mehlich, A., S. S. Bowling, and A. L. Hatfield. 1976. Buffer pH acidity in relation to nature of soil acidity and expression of lime requirement. *Commun. Soil Sci. Plant Anal.* 7:253–263.

Mikkelsen, R. L. and J. J. Camberato. 1995. Potassium, sulfur, lime, and micronutrient fertilizers, In: *Soil Amendments and Environmental Quality,* J. E. Rechcigl (ed.). Boca Raton, FL: Lewis Publishers, 109–137.

Peech, M. 1965. Hydrogen-ion activity, In: *Methods of Soil Analysis,* Part 2, Agronomy No. 9. C. A. Black (ed.). Madison, WI: American Society of Agronomy 914–926.

Reid, K. (ed.). 1998. *Soil Fertility Manual.* Toronto: Ontario Ministry of Agriculture, Food and Rural Affairs, Queen's Printer for Ontario.

Schofield, R. K. and A. W. Taylor. 1955. The measurement of soil pH. *Soil Sci. Soc. Amer. Proc.* 19:164–167.

Schwertmann, U. and M. L. Jackson. 1964. Influence of hydroxyaluminum ions on pK titration curves of hydronium-aluminum clays. *Soil Sci. Soc. Amer. Proc.* 28:179–182.

Sen Tran, T. and W. van Lierop. 1993. Lime requirement, In: *Soil Sampling and Methods of Analysis,* M. R. Carter (ed.). Boca Raton, FL: Lewis (CRC) Press, 109–113.

Shoemaker, H. E., E. O. McLean, and P. F. Pratt. 1962. Buffer methods for determination of lime requirement of soils with appreciable amount of exchangeable aluminum. *Soil Sci. Soc. Amer. Proc.* 25:274–277.

Sims, J. T. 1996. Lime requirement, In: *Methods of Soil Analysis, Part 3, Chemical Methods,* SSSA Book Series No. 5, R. L. Sparks (ed.). Madison, WI: Soil Science Society of America, 491–515.

Sims, J. T. and D. Eckert. 1995. Recommended soil pH and lime requirement tests, In: *Recommended Soil Testing Procedures for the Northeastern United States,* 2nd ed., Northeast Regional Publications No. 493, J. T. Sims and A. M. Wolf (eds.). Newark, DE: Agricultural Experiment Station, University of Delaware, 16–21.

Slattery, W. J., M. K. Conyers, and R. L. Aitken. 1999. Soil pH, aluminium, manganese, and lime requirement, In: *Soil Analysis: An Interpretation Manual,* K. I. Peverill, L. A. Sparrow, and D. J. Reuter (eds.), Collingwood, Australia: CSIRO Publishing, 103–128. Australia.

Thomas, G. W. 1982. Exchangeable cations, In: *Methods of Soil Analysis,* Part 2, 2nd ed., Agronomy No. 9. A. L. Page, R. H. Miller, and D. R. Keeney (eds.).Madison, WI: American Society of America, 159–165.

Thomas, G. W. 1996. Soil pH and soil acidity, In: *Methods of Soil Analysis, Part 3, Chemical Methods.* SSSA Book Series No. 5, D. L. Sparks (ed.). Madison, WI: Soil Science Society of America, 475–490.

van Lierop, W. 1983. Lime requirement determination of acid organic soils using buffer-pH methods. *Can. J. Soil Sci.* 63:411–423.

van Lierop, W. 1990. Soil pH and lime requirement determinations, In: *Soil Testing and Plant Analysis,* SSSA Book Series No. 3. R. L. Westerman (ed.). Madison, WI: Soil Science Society of America, 73–126.

Watson, M. E. and J. R. Brown. 1998. pH and lime requirement tests, In: *Recommended Chemical Soil Test Procedures for the North Central Region.* North Central Regional Publication No. 221 (revised), J. R. Brown (ed.). Columbia, MO: Missouri Agricultural Experiment Station SB 1001, University of Missouri.

Yuan, T. L. 1974. A double buffer method for the determination of lime requirement of acid soils. *Soil Sci. Soc. Amer. Proc.* 38:437–440.

Yuan, T. L. 1975. Lime requirement determination of sandy soils by different rapid methods. *Soil Crop Sci. Soc. Fla. Proc.* 25:274–277.

5 Conductance, Soluble Salts, and Sodicity

The level of soluble salts ("salt" in this context refers to any soluble ion) in the soil solution can have a significant impact on plant growth (Dunkle and Merkle, 1944; Richards, 1969; Maynard and Hochmuth, 1997; Ludwick, 1997; Shaw, 1999), whether the "salt level" is due to natural conditions, such as naturally occurring saline or sodic soils (Richards, 1969), the use of salt-containing irrigation water, or from the result of a heavy application of fertilizer in a small volume of soil. Salinity is also a significant factor when plants are grown in containers or growing beds on soilless media (Merkle and Dunkle, 1944; Kuehny and Morales, 1998 (See Chapter 14).

The normal procedure for measuring the soluble salt level of a soil is by extracting a portion of soil with water and then measuring the electrical conductivity (EC) of the extract using a conductivity cell and meter (Bower and Wilcox, 1965; Anon., 1983; Janzen, 1993; Anon., 1994; Gartley, 1995; Rhoades, 1996; Whitney, 1998; Shaw, 1999). To identify the ions in the extract contributing to the EC, it may be subjected to chemical analysis. In recent years, there has been a major change in the units used to express the EC of an extractant from either parts per million (ppm) or mhos (the reverse of "ohms") per centimeter (mhos cm^{-1}) to international units (SI) units, siemens (S) per meter (S m^{-1}) or decisiemens (dS) per centimeter (dS cm^{-1}). Conversion factors are given at the end of this chapter. Conductivity meters, particularly older instruments, have varied in their units of expression, but in more recent years, newer meters record in SI units.

The soil to water extract ratio must be known when interpretating an EC measurement in order to relate the soluble salt reading to expected crop response. In addition, soil type will affect an interpretation of an EC reading. It should also be remembered that crops vary widely in their tolerance to soil salinity as determined by an EC measurement (Ludwick, 1997; Maynard and Hochmuth, 1997; Shaw, 1999). The interpretative tables given in this chapter are generalized in terms of an EC measurement effect on plant performance.

The extraction and measurement procedures for determining the EC of a soil given in this chapter are those commonly used in laboratories in the U.S.

5.1 DETERMINATION OF SPECIFIC CONDUCTANCE IN SUPERNATANT 1:2 SOIL:WATER SOLUTION

1. Principle of the Method

1.1. Although specific conductance measurements in saline soils are principally carried out on a soil-paste extract, research workers in the humid soil region make extensive use of a 1:2 soil:water extraction of greenhouse soils. Specific conductance values in the 1:2 extract have been observed not to be comparable with those in the saturation extract. However, Jackson (1958) concluded that specific conductance ranges of the widely contrasting alkaline and humid regions are quite similar.

1.2. The specific conductance method described is based on the experience of the Agronomic Division, North Carolina Department of Agriculture. All measurements were made on greenhouse soils and on field problem soils. The 1:2 soil:water ratio in the procedure is based on a soil volume rather than on a soil weight basis. This avoids the need for further dilution of low bulk densities for histosols. Guidelines for restoring fields flooded by salt water are included.

2. Range and Sensitivity

2.1. The method is adapted to a wide range of salt concentrations, depending on the measuring instrument. The range can be extended by suitable dilution of the extract.

3. Interferences

3.1. Specific conductance increases with increasing temperature; hence, compensation of temperature differences from the calibrated standard is required.

3.2. In order to obtain reproducible results, clean and well-platinized electrodes are essential.

4. Precision and Accuracy

4.1. Conductivity values of less than one (<1.0) should be reported to two decimal places and values greater than one (>1.0) to three significant figures.

5. Apparatus

5.1. No. 10 (2-mm opening) sieve.
5.2. 50–60 mL cup (glass, plastic, or waxed cup of similar size).
5.3. 10-cm^3 scoop, volumetric.
5.4. 20-mL pipette.
5.5. Conductivity meter.
5.6. Conductivity cell, pipette type, 2- to 3-mL capacity.
5.7. Thermometer, 1–100°C (32–212°F).

6. Reagents (use reagent grade chemicals and pure water)

6.1. *Potassium Chloride* (0.01 N): Weigh 0.7456 g potassium chloride (KCl) into 500 mL water in a 1,000-mL volumetric flask and bring to volume with water. This solution has a conductivity of 1.41 dS m^{-1} (mmhos cm^{-1}) at 25°C (77°F).

7. Calibration and Standards

7.1. To determine the cell constant (q), use the 0.01 N KCl (see 6.1) solution at 25°C (77°F) which will will have a specific conductance (SC) of 0.0014118 dS m^{-1}.

Conductance, Soluble Salts, and Sodicity

7.2. The cell constant (q) for any commercially prepared conductivity cell can be calculated, according to Willard et al. (1968, page 720), by the following relationship:

$$K = (1/R)(d/A) = q/R$$

where: K = specific conductance, A = electrode area, d = plate spacement, and R = resistance in ohms per cm. In the case of 0.01 N KCl (see 7.1), the cell constant $(q) = 0.0014118$ (mho per cm) × R (ohms per cm). $R = 708.32$ ohm if the cell has electrodes 1 cm² in area spaced 1 cm apart (*Note:* mhos = 1 ohm).

7.3. Some conductivity instruments read in specific conductance (SC) expressed in mhos × 10^{-5} as well as resistance (ohms). Before accepting the mhos × 10^{-5} dial readings, the cell constant should be determined and the mhos × 10^{-5} dial readings verified for the cell constant used.

8. Procedure

8.1. Scoop 10 cm³ (see 5.3) 2-mm sieved soil into a beaker (see 5.5), add 20 mL water, and stir thoroughly. Allow the suspension to settle for at least 30 minutes or long enough for the solids to settle.

8.2. Draw the supernatant into the conductivity pipette to slightly above the constricted part of the pipette (see 5.6). Avoid drawing the liquid into the rubber bulb. If this occurs, rinse the bulb before continuing with the next sample.

9. Calculations

9.1. Specific conductance (*SC*) of the soil extract is calculated as follows:

$$SC \text{ [mhos per cm at } 25°C \text{ (77°F)]} = (0.0014118 \times R_{std})/R_{ext}$$

where: the value of 0.0014118 is the specific conductance (SC) of the standard 0.01 N KCl solution in dS m^{-1} at 25°C (77°F) and R_{std} and R_{ext} refer to resistance in ohms of the standard (0.01 N KCl) solution (see 6.1) and extract (see 8.1), respectively. Multiply the results by 1,000 to obtain dS m^{-1} at 25°C (77°F).

9.2. *Alternate method of calculation.* After the cell constant (q) has been determined (see 7.2), the specific conductance of the soil extract can be obtained from the following relationship:

$$SC \text{ [dS m}^{-1} \text{ at } 25°C \text{ (77°F)]} = q/R$$

where: q = determined cell constant and R = resistance in ohms of the soil extract.

10. Interpretation

10.1. Results with various soils and crops using the 1:2 soil:water ratio extraction have been reported by Dunkle and Merkle (1944) and Merkle and Dunkle (1944). Jackson (1958) has summarized the relationships of specific conductance in

1:2 soil:water extract (observed) to that in the saturation extract (calculated) for a silt loam soil at 40% saturation, and a clay loam high in organic matter soil at 100% saturation. The conductance ratios of the 1:2 saturation extract values of the 40% and 100% saturated soils were 0.2 and 0.5, respectively.

10.2. Using the 0.2 ratio values in relation to the Scofield Salinity Scale (Richards, 1969), together with published (Reisenauer, 1978) and local experience (Agronomic Division, North Carolina Department of Agriculture), a general guide to plant effects associated with different ranges of specific conductance measured in a 1:2 soil:water ratio by volume is as follows:

dS m^{-1} at 25°C	Effects
<0.40	*Nonsaline:* Salinity effects mostly negligible, excepting possibly bean and carrot.
0.40–0.80	*Very Slightly Saline:* Yield of salt-sensitive crops, such as flax, clover (alsike, red), carrot, onion, bell pepper, lettuce, and sweet potato, may be reduced by 25% to 50%.
0.81–1.20	*Moderately Saline:* Yield of salt-sensitive crops restricted. Seedlings may be injured. Satisfactory for well-drained greenhouse soils. Crop yields reduced by 25% to 50% may include broccoli and potato plus the other plants listed above.
1.21–1.60	*Saline:* Crops tolerant include cotton, alfalfa, cereals, grain sorghums, sugar beet, Bermuda grass, tall wheat grass, and Harding grass. Salinity higher than desirable for greenhouse soils.
1.61–3.20	*Strongly Saline:* Only salt-tolerant crops yield satisfactory. For greenhouse crops, leach soil with enough water so that two to four quarts pass through each square foot of bench area or one pint of water per six-inch pot; repeat after about one hour. Repeat again if readings are still in the high range.
>3.20	*Very Strongly Saline:* Only salt-tolerant grasses, herbaceous plants, certain shrubs and trees will grow.

11. Guidelines for Restoring Fields Flooded by Salt water

11.1. Agricultural extension specialists at North Carolina State University and agronomists of the Agronomic Division, North Carolina Department of Agriculture, suggest that the following procedures be used when cropland is flooded by salt water or salt spray is blown inland by hurricane-force winds (see 11.6).

11.2. Plants growing on saltwater flooded soil exhibit the greatest damage when soil moisture is limiting growth. When the soil is relatively dry, the salt concentration of the soil solution around the plant roots is the highest and prevents uptake of moisture. On the other hand, saltwater may wash across fields doing little or no damage if the soil has been previously saturated by rain or fresh-water floods.

11.3. Treatment for returning land to a productive level is based on the salt content from properly collected samples of the suspected salt-damaged area. Collect core samples to a depth of 18 inches (45.7 cm) from each field. Divide cores into four parts as follows: (a) 0–2 inches (0–5 cm); (b) 2–6 inches (5–15 cm); (c) 6–12 inches

Conductance, Soluble Salts, and Sodicity

(15–30 cm) and (d) 12–18 inches (30–46 cm). Salt concentration is determined on a 1:2 soil:water sample and the results expressed in dS m^{-1}.

11.4. The effects and reclamation are as follows:

dS m^{-1}	Effects and Reclamation
< 0.4	Most crops will grow quite well; no injury should be expected.
0.4–0.8	Fairly safe for most crops; however, a long dry spell may draw salts up near surface and damage plants.
0.81–1.2	Only salt-tolerant crops as listed below will grow. Reclamation for other crops as suggested below.
> 1.2	Few crops will survive; reclamation necessary.

11.5. Crops Tolerance to Salt

11.5.1. *Tolerant:* Rape, kale, cotton, barley, tall fescue, garden pea, Rhodes grass, and Bermuda grass. Turf grasses and ornamentals, i.e., zoysia, St. Augustine, American beachgrass, sea oat, English ivy, dwarf yaupon, several species of yucca, dwarf natal-plum, sea grape, Japanese privet, common oleander, and wax myrtle.

11.5.2. *Moderately tolerant:* Fig, grape, wheat, oat, rye, sunflower, corn, ryegrass, alfalfa, sweet clover, Sudan grass, birdsfoot trefoil, orchardgrass, carrot, lettuce, onion, and tomato.

11.5.3. *Sensitive:* Pear, peach, apple, plum, vetch, field beans, green beans, red clover, white clover, alsike clover, ladino clover, cabbage, potato, and many others.

11.6. Reclamation

11.6.1. *Procedure:* Apply calcium sulfate (landplaster, gypsum) (CaSO$_4$) to the fields if dS m^{-1} is above 1.00 in the top six inches of the soil. The application rate is as follows: (a) less than 2% organic matter, 2,000 lbs A^{-1}; (b) 2% to 5% organic matter, 3,000 pounds A^{-1}; and (c) above 5% organic matter, 4,000 pounds A^{-1}.

11.6.2 Resample in three to six months to determine the progress of treatments. Since calcium sulfate (CaSO$_4$) contributes to the specific conductance, it is essential to determine calcium (Ca) and sodium (Na) in the extract also.

5.2 DETERMINATION OF SPECIFIC CONDUCTANCE IN SUPERNATANT 1:1 SOIL:WATER SOLUTION

1. Principle of the Method

1.1. The electrical conductivity (EC) value obtained by 1:1 soil-to-water method is not as easily interpreted as that for the saturated paste method. Interpretation given is based on soil texture.

2. Range and Sensitivity

2.1. The method is adapted to a wide range of salt concentrations, depending on the measuring instrument. The range can be extended by suitable dilution of the extract.

3. Interferences

3.1. Specific conductance increases with increasing temperature; hence, compensation of temperature differences from the calibrated standard is required.
3.2. In order to obtain reproducible results, clean and well-platinized electrodes are essential.

4. Precision and Accuracy

4.1. Conductivity values of less than one (<1.0) should be reported to two decimal places and values more than one (>1.0) to three significant figures.

5. Apparatus

5.1. No. 10 (2-mm opening) sieve.
5.2. 50–60 mL cup (glass, plastic, or waxed cup of similar size).
5.3. 10-cm^3 scoop, volumetric.
5.4. 20-mL pipette.
5.5. Conductivity meter.
5.6. Conductivity cell, pipette type, 2- to 3-mL capacity.
5.7. Thermometer, 1–100°C (32–212°F).

6. Reagents (use reagent grade chemicals and pure water)

6.1. *Potassium Chloride* (0.01 N): Weigh 0.7456 g potassium chloride (KCl) into 500 mL water in a 1,000-mL volumetric flask and bring to volume with water. This solution has a conductivity of 1.41 dS m^{-1} (mmhos cm^{-1}) at 25°C (77°F).

7. Calibration and Standards

7.1. To determine the cell constant (q), use the 0.01 N KCl (see 6.1) solution at 25°C (77°F) which will will have a specific conductance (SC) of 0.0014118 dS m^{-1}.
7.2. The cell constant (q) for any commercially prepared conductivity cell can be calculated, according to Willard et al. (1968, page 720), by the relationship:

$$K = (1/R)(d/A) = q/R$$

where: K = specific conductance, A = electrode area, d = plate spacement, and R = resistance in ohms per cm. In the case of 0.01N KCl (see 7.1), the cell constant (q) = 0.0014118 (mho per cm) × R (ohms per cm). R = 708.32 ohm if the cell has electrodes 1 cm^2 in area spaced 1 cm apart (*Note:* mhos = 1 ohm).
7.3. Some conductivity instruments read in specific conductance (SC) are expressed in mhos × 10^{-5} as well as resistance (ohms). Before accepting the mhos × 10^{-5} dial

Conductance, Soluble Salts, and Sodicity

readings, the cell constant should be determined and the mhos × 10⁻⁵ dial readings verified for the cell constant used.

8. Procedure

8.1. Scoop 20 cm³ (see 5.3) 2-mm sieved soil into a test tube or small container (see 5.5), add 20 mL water, and stir thoroughly. Allow the suspension to stand for 15 to 20 minutes.

8.2. Insert the conductivity cell into the suspension and read the electrical conductivity (EC).

9. Calculations

9.1. Specific conductance *(SC)* of the soil extract is calculated as follows:

$$SC \text{ [mhos per cm at 25°C (77°F)]} = (0.0014118 \times R_{std})/R_{ext}$$

where: the value of 0.0014118 is the specific conductance *(SC)* of the standard 0.01 N KCl solution in dS m⁻¹ at 25°C (77°F) and R_{std} and R_{ext} refer to resistance in ohms of the standard (0.01 N KCl) solution (see 6.1) and extract (see 8.1), respectively. Multiply the results by 1,000 to obtain dS m⁻¹ at 25°C (77°F).

9.2. *Alternate Method of Calculation:* After the cell constant **(q)** has been determined (see 7.2), the specific conductance of the soil extract can be obtained from the following relationship:

$$SC \text{ [dS m}^{-1} \text{ at 25°C (77°F)]} = q/R$$

where: q = determined cell constant and R = resistance in ohms of the soil extract.

10. Interpretation

Relationship between Conductivity (EC) and Degree of Salinity by the 1:1 Method

Degree of Salinity	Course to Loamy Sand	Loamy Fine Sand to Loam	Silt Loam to Clay Loam	Silty Clay Loam to Clay
		dS m⁻¹		
Non-Saline	0–1.1	0.1.2	0–1.31	0–1.4
Slightly Saline	1.2–2.4	1.3–2.4	1.4–2.5	1.5–2.8
Moderately Saline	2.5–4.4	2.5–4.7	2.6–5.0	2.9–5.7
Strongly Saline	4.5–8.9	4.8–9.4	5.1–10.0	5.8–11.4
Very Strongly Saline	>9.0	>9.5	>10.1	>11.5

5.3 DETERMINATION OF SPECIFIC CONDUCTANCE BY SATURATED PASTE METHOD

1. Principle of the Method

1.1. The saturated paste method has been the long-term procedure for measuring soil salinity for assessing its affect on plant growth. No adjustment due to soil texture is required.

2. Range and Sensitivity

2.1. The method is adapted to a wide range of salt concentrations, depending on the measuring instrument. The range can be extended by suitable dilution of the extract.

3. Interferences

3.1. Specific conductance increases with increasing temperature; hence, compensation of temperature differences from the calibrated standard is required.
3.2. In order to obtain reproducible results, clean and well-platinized electrodes are essential.

4. Precision and Accuracy

4.1. Conductivity values of less than one (<1.0) should be reported to two decimal places and values more than one (>1.0) to three significant figures.

5. Apparatus

5.1. No. 10 (2-mm opening) sieve.
5.2. 400-mL beaker.
5.3. Spatula.
5.4. Filter funnel.
5.5. Vacuum pump.
5.6. Conductivity meter.
5.7. Conductivity cell, pipette type, 2- to 3-mL capacity.
5.8. Thermometer, 1–100°C (32–212°F).
5.9. Analytical balance.

6. Reagents (use reagent grade chemicals and pure water)

6.1. *Potassium Chloride* (0.01 N): Weigh 0.7456 g potassium chloride (KCl) into 500 mL water in a 1,000-mL volumetric flask and bring to volume with water. This solution has a conductivity of 1.41 dS m^{-1} (mmhos cm^{-1}) at 25°C (77°F).

7. Calibration and Standards

7.1. To determine the cell constant (q), use the 0.01 N KCl (see 6.1) solution at 25°C (77°F) which will will have a specific conductance (SC) of 0.0014118 dS m^{-1}.

Conductance, Soluble Salts, and Sodicity

7.2. The cell constant (q) for any commercially prepared conductivity cell can be calculated, according to Willard et al. (1968, page 720), by the relationship:

$$K = (1/R)(d/A) = q/R$$

where: K = specific conductance, A = electrode area, d = plate spacement, and R = resistance in ohms per cm. In the case of 0.01 N KCl (see 7.1), the cell constant $(q) = 0.0014118$ (mho per cm) × R (ohms per cm). $R = 708.32$ ohm if the cell has electrodes 1 cm^2 in area spaced 1 cm apart (*Note:* mhos = 1 ohm).

7.3. Some conductivity instruments read in specific conductance (SC) are expressed in mhos × 10^{-5} as well as resistance (ohms). Before accepting the mhos × 10^{-5} dial readings, the cell constant should be determined and the mhos × 10^{-5} dial readings verified for the cell constant used.

8. Procedure

8.1. Weigh 250 g 2-mm sieved soil (see 5.1) into a 400-mL beaker (see 5.2) and add water while stirring with a spatula (see 5.3) until the soil slides freely from the surface of the spatula thoroughly. At saturation, the soil paste will glisten as its reflects light. Let stand for one hour and recheck for saturation.

8.2. Transfer the saturated paste to a filter funnel (see 5.4) and draw water from the saturated soil by applying vacuum (see 5.5). Collect the filtrate and determine its conductivity.

9. Calculations

9.1. Specific conductance *(SC)* of the soil extract is calculated as follows:

$$SC\,[\text{mhos per cm at 25°C (77°F)}] = \mathbf{(0.0014118 \times R_{std})/R_{ext}}$$

where: the value of 0.0014118 is the specific conductance (SC) of the standard 0.01 N KCl solution in dS m^{-1} at 25°C (77°F) and R_{std} and R_{ext} refer to resistance in ohms of the standard (0.01 N KCl) solution (see 6.1) and extract (see 8.1), respectively. Multiply the results by 1,000 to obtain dS m^{-1} at 25°C (77°F). Report specific conductance (SC) values in mmho per cm.

9.2. *Alternate Method of Calculation:* After the cell constant *(q)* has been determined (see 7.2), the specific conductance of the soil extract can be obtained from the following relationship:

$$SC\,[\text{dS m}^{-1}\text{ at 25°C (77°F)}] = q/R$$

where: q = determined cell constant and R = resistance in ohms of the soil extract.

5. Interpretation

Relationship between Conductivity (EC) and Degree of Salinity by the Saturation Method for All Soils

Degree of Salinity	dS m^{-1}
Nonsaline	0.0–2.0
Slightly Saline	2.1–4.0
Moderately Saline	4.1–8.0
Strongly Saline	8.1–16.0
Very Strongly Saline	>16.1

Electrical Conductivity Units and Conversions

1 millimho per centimeter (mmho cm^{-1}) = EC 10^{-3}
1 micromho per centimeter (mmho cm^{-1}) = EC × 10$^-$
1,000 micromhos per centimeter = 1 mmho cm^{-1}
1 millisiemen per centimeter (1 mS cm^{-1}) = 1 μmho cm^{-1}
1 decisiemen per meter (1 dS m^{-1}) = 1 mmho cm^{-1} = 700 ppm

5.4 REFERENCES

Anon. 1983. Determination of specific conductance in supernatant 1:2 soil:water solution, In: *Reference Soil Test Methods for the Southern Region of the United States,* Southern Cooperative Series Bulletin 289, R.A. Isaac (ed.). Athens, GA: The University of Georgia College of Agriculture Experiment Stations, 8–14.

Anon. 1994. Soluble salts, in: *Plant, Soil, and Water Reference Methods for the Western Region,* A Western Regional Extension Publication WREP 125, R. G. Galvak, D. A. Horneck, and R. O. Miller (eds.). Fairbanks, AK: University of Alaska, 38.

Bower, C. A., and L. U. Wilcox. 1965. Soluble salts, In: *Methods of Soil Analysis, Part 2,* Agronomy No. 9, C. A. Black (ed.). Madison, WI: American Society of Agronomy, 938–951.

Dunkle, E. C. and F. G. Merkle. 1944. The conductivity of soil extraction in relation to germination and growth of certain plants. *Soil Sci. Soc. Amer. Proc.* 8:185–188.

Gartley, K. 1995. Recommended soluble salts tests, In: *Recommended Soil Testing Procedures for the Northeastern United States.* Second Edition. Northeast Regional Publication No. 493 (revised), J. T. Sims and A. M. Wolf (eds.). Newark, DE: Agricultural Experiment Station, University of Delaware, 70–75.

Jackson, M. L. 1958. *Soil Chemical Analysis.* Englewood Cliffs, NJ: Prentice-Hall.

Janzen, H. H. 1993. Soluble salts, In: *Soil Sampling and Methods of Analysis,* M. R. Carter (ed.). Boca Raton, FL: CRC Press, 161–166.

Kuehny, J. S. and B. Morales. 1998. How salinity and alkalinity levels in media affect plant growth. *Greenhouse Prod. News* 8(6):42–43.

Ludwick, A. E. (ed.). 1997. *Western Fertilizer Handbook, Second Horticultural Edition.* Danville, IL: Interstate Publishers, Inc.

Maynard, D. N. and G. J. Hochmuth. 1997. *Knott's Handbook for Vegetable Growers,* 4th ed. New York: John Wiley & Sons.

Merkle, F. G. and E. C. Dunkle. 1944. The soluble salt content of greenhouse soils as a diagnostic aid. *J. Amer. Soc. Agron.* 36:10–19.

Reisenauer, H. M. 1978. *Soil and Plant Tissue Testing in California,* Bulletin 1987 (revised ed.). Berkeley, CA: Division of Agricultural Science, University of California.

Rhoades, J. D. 1996. Salinity: Electrical conductivity and total dissolved solids, In: *Methods for Soil Analysis, Part 3, Chemical Methods,* SSSA Book Series Number 5. R. L. Sparks (ed.). Madison, WI: Soil Science Society of America, 417–435.

Richards, L. A. (ed.). 1969. *Diagnosis and Improvement of Saline and Alkali Soils,* USDA Agricultural Handbook No. 60. Washington, DC: United States Government Printing Office.

Shaw, R. J. 1999. Soil salinity—electrical conductivity and chloride, In: *Soil Analysis: An Interpretation Manual,* K. I. Peverill, L. A. Sparrow, and D. J. Reuter (eds.). Collingwood, Australia: CSIRO Publishing, 129–145.

Willard, H. H., L. L. Merritt, Jr., and J. A. Dean. 1968. *Instrumental Methods of Analysis,* 4th ed. Princeton, NJ: D. Van Nostrand Co., Inc.

Whitney, D. A. 1998. Soil salinity, In: *Recommended Chemical Soil Test Procedures for the North Central Region,* North Central Regional Research Publication No. 221 (revised), J. R. Brown (ed.). Columbia, MO: Missouri Agricultural Experiment Station SB 1001, University of Missouri, 59–60.

6 Phosphorus

Phosphorus (P) exists in various forms in mineral soils, being about equally divided between soil organic matter and various inorganic forms. The latter P forms are primarily mixtures of aluminum, iron, and calcium phosphates. The relative percentage among these three forms is a function of soil pH and the higher percentages of aluminum and iron phosphates occurring in acid soils, while a higher percentage as calcium phosphate occurring in neutral to alkaline soils is shown in Figure 6.1. Consequently, the extraction procedure for the measurement of plant-available P will be governed to a large degree by soil pH.

Phosphorus soil and plant chemistry has been discussed by Khasawneh et al. (1980), Anon. (1996), Jones (1998), and Moody and Bolland (1999), and reviews on P soil testing procedures have been published by Olsen and Sommers (1982), Fixen and Grove (1990), Kuo (1996), and Radojevic and Bashkin (1999). Reviews on the use of P fertilizers can be found in Black (1993), Withers and Sharpley (1995), and P fertilizer recommendations for some crops in the books by Maynard and Hochmuth (1997) and Ludwick (1998). Comparisons among P extraction procedures are given by Hanlon and Johnson (1984), Wolf and Baker (1985), Sims (1989), Beegle and Oravec (1990), and Schmisek et al. (1998).

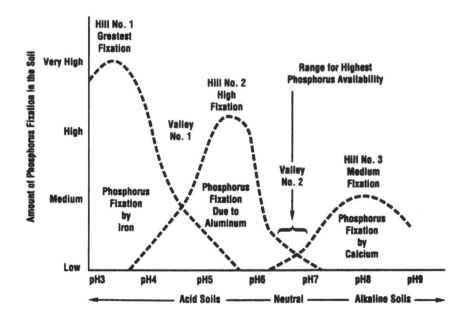

FIGURE 6.1 Availability of phosphorus varies with soil pH (Anon., 1996).

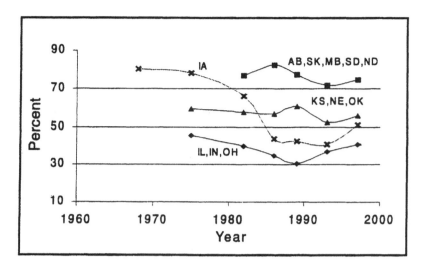

FIGURE 6.2 Percent of soil samples testing medium or lower in phosphorus (Anon., 1998).

The current soil P test level in North American soils for 1997 has been published by the Potash & Phosphate Institute (Anon., 1998), which also included a graph of the changing trend in soil P test levels from 1975 to 1996 with levels remaining fairly constant during this time period although the percentage of soils testing "medium or below" varied from a high of 80% to a low of 30% depending on state as shown in Figure 6.2.

Four soil analysis procedures for P determination are described in this chapter.

6.1 BRAY P1 EXTRACTION

1. Principle of the Method

1.1. The extraction of phosphorus (P) by the Bray P1 (Bray and Kurtz, 1945) method is based upon the solubilization effect of the hydrogen (H^+) ions on soil P and the ability of the fluoride (F^-) ion to lower the activity of aluminum (Al^{3+}) and to a lesser extent that of calcium (Ca^{2+}) and iron (Fe^{3+}) ions in the extraction system. As described in this section, clay soils with a moderately high degree of base saturation or silty clay loam soils that are calcareous or have a very high degree of base saturation will lessen the solubilizing ability of the extractant. Consequently, the method should normally be limited to soils with pH_w values less than 6.8 when the texture is silty clay loam or finer.

Calcareous soils, or high pH, fine textured soils may be tested by this method, but higher ratios of extractant-to-soil are often used for such soils (Smith et al., 1957). Other alternatives are the Olsen P (Olsen et al., 1954) and Mehlich No. 3 (Mehlich, 1984) extractant procedures, although the Mehlich No.3 extract is not widely accepted for such use. The Bray P1 method is also suitable for organic soils.

1.2. The extractant was developed and first described by Bray and Kurtz (1954). The extraction time and the solution-to-soil ratio in their procedure were 40 seconds and 7-mL extractant to 1.0 g soil, respectively. To simplify adaptation to routine laboratory work and to extend the range of soils for which the extractant is suitable, both the extraction time and the solution to soil ratio have been altered to five minutes and a 1:10 soil:extractant ratio. This modification is in wide use in laboratories of the mideast, midsouth and north central areas of the United States (Jones, 1973, 1998). This method is described by Anon. (1983a, 1994c) and Frank et al. (1998).

1.3. Comparisons have been made between Mehlich No. 1 and other procedures by Schmisek et al. (1989), Sims (1989), and Wolf and Baker (1989).

2. Range, Sensitivity, and Methods of Analysis

2.1. This procedure yields a standard curve that is essentially linear to 10 mg P L^{-1} in the soil extract (approximately 200 kg ha^{-1} or 178 lb A^{-1} of extractable P).

2.2. The sensitivity is approximately 0.15 mg P L^{-1} in the extract (2.7 lb P A^{-1} or 3.0 kg P ha^{-1} in the soil).

2.3. The commonly used method of analysis is UV-VIS spectrophotometry (Watson and Isaac, 1990; Wright and Stuczynski, 1996).

3. Interferences

3.1. *Arsenic:* Concentrations of up to 1 mg arsenic (As) L^{-1} in the extract do not interfere (Murphy and Riley, 1962). Jackson describes techniques for removal (Jackson, 1958).

3.2. *Silicon:* Silicon (Si) will not interfere at < 10 mg Si L^{-1} in the extract (Murphy and Riley, 1962).

3.3. *Fluoride:* The fluoride (F) in the extract normally will not interfere in the formation of the molybdenum blue color with P. Any interference may be eliminated with the addition of boric acid (H$_3$BO$_3$) (Kurtz, 1942). Maximum color development is slower in the presence of the F$^-$ ion.

4. Precision and Accuracy

4.1. The reproducibility of determinations by this procedure depends upon the extent to which the times of extraction, filtration and color development are controlled. Reasonable control and thorough sample preparation should give a coefficient of variation of about 5%.

5. Apparatus

5.1. No. 10 (2-mm opening) sieve.
5.2. Scoop, 1.70-cm^3 volumetric.
5.3. Extraction bottle or flask, 50-mL with stoppers.
5.4. Mechanical reciprocating shaker, minimum of 180 oscillations per minute.
5.5. Filter funnel, 11 cm.
5.6. Funnel rack.
5.7. Whatman No. 2 filter paper or equivalent, 11 cm.

5.8. UV-VIS spectrophotometer set at 880 nm.
5.9. Spectrophotometer tube or cuvet.
5.10. Volumetric flasks and pipettes as required for preparation of reagents, standard solutions, and color development.
5.11. Analytical balance.

6. Reagents (use reagent grade chemicals and pure water)

6.1. *Ammonium Fluoride* (1 N): weigh 37 g ammonium fluoride (NH_4F) into a 1,000-mL volumetric flask and bring to volume with water. Store in a polyethylene container and avoid prolonged contact with glass.

6.2. *Hydrochloric Acid* (0.5 N): dilute 20.4 mL conc. hydrochloric acid (HCl) to 500 mL with water.

6.3. *Extraction Reagent:* Mix 30 mL 1 N NH_4F (see 6.1) with 50 mL 0.5 N HCl (see 6.2) in a 1,000-mL volumetric flask and dilute to volume with water. This solution is 0.03 N in NH_4F and 0.02 N in HCl and has a pH of 2.6. Store in polyethylene. This reagent is stable for more than one year.

6.4. *Ascorbic Acid Solution:* Weigh 132.0 g ascorbic acid into a 1,000 mL volumetric flask and bring to volume with water. Store in dark glass bottle in a refrigerated compartment.

6.5. *Sulfuric-Molybdate-Tartrate Solution:* Weigh 60 g ammonium molybdate [$(NH_4)_6Mo_7O_{24} \cdot 4H_2O$] into a 1,000-mL volumetric flask and add 500 mL water. Add 1.455 g antimony potassium tartrate [$K(SbO)C_4H_4O_6 \cdot 1/2H_2O$] to the molybdate solution. Slowly add 700 mL conc. sulfuric acid (H_2SO_4) and mix well. Let cool and dilute to 1,000 mL with water. This solution may be blue but will produce a clear solution when the working solution (see 6.6) is prepared. Store in a polyethylene or Pyrex bottle in a dark, refrigerated compartment.

6.6. *Working Solution:* Pipette 10 mL Ascorbic Acid Solution (see 6.4) into a 1,000-mL volumetric flask and add about 800 mL water followed by 25 mL Sulfuric-Molybdate-Tartrate Solution (see 6.5), and then dilute to 1,000 mL with water. Allow to stand at least one hour before using. Prepare fresh daily.

6.7. *Phosphorus Standard* (100 mg P L^{-1}): Commercially prepared standard, or weigh 0.4394 g monobasic potassium phosphate (KH_2PO_4) that has been oven-dried at 100°C into a 1,000-mL volumetric flask and bring to volume with Extraction Reagent (see 6.3).

7. Procedure

7.1. *Extraction:* Weigh 2.0 g or scoop 1.70 cm^3 air-dry < 10-mesh (2-mm) soil into a 50-mL extraction bottle or flask (see 5.3), add 20 mL Extraction Reagent (see 6.3) and shake for five minutes on a reciprocating shaker (see 5.4). Filter through Whatman No. 2 filter paper (see 5.6), limiting the filtration time to 10 minutes, and save the extract.

7.2. *Color Development:* Transfer exactly 2.0 mL extract (see 7.1) and P standards (see 8.1) into a spectrophotometer tube or cuvet (see 5.9), add 8 mL Working Solution (see

Phosphorus

6.6), and mix the contents of the tube thoroughly. After 10 minutes, measure the percentage of transmittance (%T) at 882 nm. The color intensity is stable for four hours.

8. Calibration and Standards

8.1. *Working Phosphorus Standards:* Using the Phosphorus Standard (100 mg P L^{-1}) (see 6.7), prepare six Working Standards containing from 0.2 to 10 mg P L^{-1} in the final volume. Make all dilutions with Extraction Reagent (see 6.3).

8.2. *Calibration Curve:* On semilog graph paper, plot the percentage of transmittance (%T) on the logarithmic scale versus mg P L^{-1} in the Working Standards on the linear scale.

9. Calculation

9.1. The results are reported as kg P ha^{-1} for a 20 cm depth of soil: kg P ha^{-1} in the soil = mg P L^{-1} in the extractant $\times$ 22.4, lbs P A^{-1} = mg P L^{-1} in the extractant $\times$ 20 (*Note:* assumes that a uniform 2.0 mL aliquot is used for standards and unknowns in 7.2).

10. Effects of Storage

10.1. After air drying, the extractable P levels in soils remain stable for several months. The influence of storage time and temperature is discussed by Houba and Novozamsky (1998).

10.2. After extraction, the P in the extract should be measured within 72 hours.

11. Interpretation

11.1. Accurate fertilizer recommendations for P must be based on field response data conducted under local soil-climate-crop conditions (Peck et al., 1977; Brown, 1987; Dahnke and Olson, 1990; Withers and Sharpley, 1995). In general, the extractable P levels may be categorized as follows:

Category	Extractable P in Soil kg ha^{-1}	lb A^{-1}
Low	<34	<30
Medium	34–68	30–60
High	> 68	>60

6.2 OLSEN'S SODIUM BICARBONATE EXTRACTION

1. Principle of the Method

1.1. The extractant is a 0.5 M sodium bicarbonate ($NaHCO_3$) solution at a pH of 8.5 (Olsen et al., 1954). The solubility of calcium phosphate in calcareous, alkaline, or neutral soils is increased because of the precipitation of calcium (Ca^{2+}) as calcium carbonate ($CaCO_3$). In acid soils, phosphorus (P) concentration in solution increases when aluminum and iron phosphates, such as variscite and strengite, are present (Lindsay and Moreno, 1960). Secondary precipitation reactions are reduced in acid and calcareous soils because iron (Fe), aluminum (Al), and Ca concentrations remain low in the extract (Olsen and Dean, 1965). Recent studies have shown that the maintenance of the pH of the extractant at 8.5 is essential to obtain reliable results.

1.2. The extractant was first developed and described by Olsen et al. (1954). The original procedure required that 5 g soil be shaken for 30 minutes in 100 mL extraction reagent containing 1 teaspoon of carbon black (Darco G-60). The use of carbon black eliminated the color in the extract. This procedure was later modified so that the use of carbon black was eliminated (Watanabe and Olsen, 1965). In the modified method, a single solution reagent which consists of an acidified solution of ammonium molybdate containing ascorbic acid and a small amount of antimony (Sb) is used (Murphy and Riley, 1962; Watanabe and Olsen, 1965). The method is described by Anon. (1994b) and Frank et al. (1998). Comparison between the Olsen and other methods of P extraction has been made by Wolf and Baker (1989).

2. Range, Sensitivity, and Methods of Analysis

2.1. The molybdate complex obeys Beer's Law, yielding a straight line when the log of % transmittance versus concentration is plotted up to a concentration of 2 mg P L^{-1} in the final solution (Murphy and Riley, 1962).

2.2. The sensitivity of this method is 0.02 mg P L^{-1} in the extract (Watanabe and Olsen, 1965).

2.3. The commonly used method of analysis is UV-VIS spectrophotometry (Watson and Isaac, 1990; Wright and Stuczynski, 1996).

3. Interferences

3.1. This method provides noninterference of silicate in solution up to at least 10 mg silicon (Si) L^{-1}, up to at least 50 mg Fe(III) L^{-1}, up to at least 10 mg copper [Cu(II)] L^{-1}, and up to at least 1 mg arsenate L^{-1} (Murphy and Riley, 1962).

4. Precision

4.1. The coefficient of variation (CV) depends on the concentration of P in the extract. For routine analysis, the CV may vary from 10 to 20%.

5. Apparatus

5.1. No. 10 (2-mm opening) sieve.
5.2. 2-cm^3 scoop, volumetric.
5.3. 250-mL extraction bottle or flask, with stoppers.
5.4. Mechanical reciprocating shaker, 180 oscillations per minute.
5.5. Filter funnel, 11 cm.
5.6. Funnel rack.
5.7. Whatman No. 40 filter paper (or equivalent), 12.5 cm.
5.8. UV-VIS spectrophotometer set at 880 nm.

Phosphorus

5.9. Spectrophotometric tube or cuvet.
5.10. Analytical balance.
5.11. Volumetric flasks and pipettes as required for preparation of reagents, standard solutions, and color development.

6. Reagents (use reagent grade chemicals and pure water)

6.1. *Extraction Reagent* (0.5 N Sodium Bicarbonate): Dissolve 42.0 g sodium bicarbonate ($NaHCO_3$) in water in a 1,000-mL volumetric flask and bring to Volume with water. Adjust the pH to 8.5 with 50% sodium hydroxide (NaOH) or 0.5 N hydrochloric acid (HCl). Add mineral oil to avoid exposure of the solution to air. Store in a polyethylene container and check the pH of the solution before use and adjust if necessary (*Note:* maintenance of the pH at 8.5 is essential).

6.2. *Mixed Reagent:* Weigh 12.0 g ammonium molybdate $[(NH_4)_6Mo_7O_{24}\cdot 4H_2O]$ into 250 mL water, and mix to dissolve. Weight 0.2908 g antimony potassium tartrate $[K(SbO)C_4H_4O_6\cdot 1/2\ H_2O]$ into 500 mL 5N sulfuric acid (H_2SO_4) (148 mL conc. $H_2SO_4\ L^{-1}$), and mix to dissolve. Mix the two solutions together thoroughly in a 2,000 mL volumetric flask and bring to volume with water. Store in a Pyrex bottle in a dark, cool place.

6.3. *Color-Developing Reagent:* Add 0.739 g ascorbic acid to 140 mL Mixed Reagent (see 6.2). This amount of reagent is enough for 24 P determinations, allowing 20 mL for waste. This reagent should be prepared as required as it will not keep for more than 24 hours.

6.4. *Phosphorus Standard* (100 mg P L^{-1}): Commercially prepared standard, or weigh 0.4394 g monobasic potassium phosphate (KH_2PO_4) that has been oven-dried at 110°C into a 1,000-mL volumetric flask, add 200 mL water to dissolve, and bring to volume with water. Add 5 drops of toluene and shake the flask vigorously. This solution contains 100 mg P mL^{-1}.

6.5. *Working Phosphorus Standards:* Pipette 1, 2, 5, 10, 20, and 30 mL of the 100 mg P L^{-1} Phosphorus Standard (see 6.4) into 100-mL volumetric flasks and bring to volume with Extraction Reagent (see 6.1) to give 0.1, 0.2, 0.5, 1.0, 2.0, and 3.0 mg P mL^{-1} working P standards, respectively.

7. Procedure

7.1. *Extraction:* Weigh 2.5 g or scoop 2-cm^3 (see 5.2) air-dry, <10-mesh (2-mm) soil into a 250-mL extraction bottle (see 5.3), add 50 mL Extraction Reagent (see 6.1), and shake for 30 minutes on a reciprocating shaker (see 5.4). Filter and collect the filtrate (*Caution:* Soil extraction is sensitive to temperature changing 0.43 mg P kg^{-1} for each degree C for soils containing 5 to 40 mg P kg^{-1}).

7.2. *Color Development:* Pipette 5 mL extract (see 7.1) and Working Phosphorus Standards (see 6.5) into spectrophotometric tubes which are optically similar (see 5.9). Add 5 mL Color-Developing Reagent (see 6.3) carefully to prevent loss of the sample due to excessive foaming. Add 15 mL water with a burette and stir. Let it stand 15 minutes and measure the color intensity (%T) at 880 nm. If the color is too intense for the working range, reduce the aliquot from 5 mL to a lower volume and add sufficient Extraction Reagent (see 6.1) to bring the aliquot volume to 5 mL (if 1 mL aliquot is used, 4 mL Extraction Reagent is added to the aliquot). Develop the color. The color is stable for 24 hours.

8. Calibration and Standards

8.1. *Color Development of Phosphorus Standards:* Pipette aliquots of dilute Working Phosphorus Standards (see 6.5) containing from 2 to 25 mg P L^{-1} (1 mL to 12.5 mL) into 25-mL volumetric flasks. Add 5 mL Color-Developing Reagent (see 6.3) and mix. Bring to volume with the Extraction Reagent and mix thoroughly. Let stand 15 minutes and read the transmittance at 880 nm.

8.2. *Calibration Curve:* On semi-log graph paper, plot the percentage of transmittance (% T) on the logarithmic scale versus mg P L^{-1} in the solution on the linear scale. Construct a table from the calibration curve showing the relationship between the % T and P concentration. Check two points on the calibration curve on every day of P analysis.

9. Calculation

9.1. The results are reported as kg P ha^{-1} for a 20-cm depth of soil: kg P ha^{-1} in the soil = mg L^{-1} in the final solution × 250. These calculations are for a 5 mL aliquot of extract and should be adjusted for other aliquots. To express the results on a soil-weight basis, use the following formula: mg P kg^{-1} in soil = mg P L^{-1} in solution × 20.

10. Effects of Storage

10.1. Air-drying and storage may have some effect on NaHCO$_3$-extractable P. However, for routine soil testing purposes, this effect is not significant. The influence of storage time and temperature is discussed by Houba and Novozamsky (1998).

10.2. After extraction, measure the P in the extract within two hours.

11. Interpretation

11.1. It has been shown by several workers (Olsen and Dean, 1965) that a P content of <12 kg ha^{-1} in soil indicates a crop response to P fertilizers, between 12 and 24 kg ha^{-1} indicates a probable response, and >24 kg ha^{-1} indicates a crop response is unlikely. However, differences in climatic conditions and crop species may make the general guidelines given above not applicable to all conditions (Peck et al., 1977; Brown, 1987; Dahnke and Olson, 1990; Withers and Sharpley, 1995).

6.3 MEHLICH NO. 1 EXTRACTION

1. Principle of the Method

1.1 This method is primarily used to determine phosphorus (P) in sandy soils which have exchange capacities of less than 10 milliequivalents per 100 grams, are acid (pH less than 6.5) in reaction, and are relatively low (less than 5%) in organic matter content. The method is not suited for alkaline soils.

1.2 This method was first published by Mehlich (1953) and then by Nelson et al. (1953) as the North Carolina Double Acid Method. The method, which has been renamed Mehlich No. 1, is adaptable to the coastal plain soils of eastern U.S. It is currently being used by a number of state soil testing laboratories in the U.S. (Alabama, Delaware, Florida, Georgia, Maryland, New Jersey, South Carolina, and Virginia) (Jones, 1973, 1998). This method is described by Anon. (1983b), and Wolf and Beegle (1995).

2. Range, Sensitivity, and Methods of Analysis

2.1. Phosphorus can be extracted and determined in soil concentrations from 2 to 200 kg P ha^{-1} without dilution. The upper limit may be extended by diluting the extract prior to UV-VIS spectrophometric determination.

2.2. The sensitivity varies depending on the method of color development. Greater sensitivity can be obtained with the molybdophosphoric blue color method (Jackson, 1958; Watanabe and Olsen, 1965) as compared to the vanadomolybdophosphoric acid color method (Watanabe and Olsen, 1965). The estimated precision of the method is ± 1 mg P L^{-1}.

2.3. The commonly used method of analysis is UV-VIS spectrophotometry (Watson and Isaac, 1990; Wright and Stuczynski, 1996). Mehlich No. 1 being a multi-element extraction reagent, P as well as other elements can be determined by inductively coupled plasma emission spectrometry (ICP-AES) (Soltanpour et al., 1996, 1998).

3. Interferences

3.1. With some soils, the extract may be colored, varying from light to dark yellow. If the vanadomolybdophosphoric acid method (Jackson, 1965) is employed, as originally prescribed for the Mehlich No. 1 acid method (Mehlich, 1953), decolorizing is necessary to avoid obtaining high results. Decolorization can be accomplished by including *activated* charcoal in the extraction procedure. Decolorization is not necessary if color development is by the molybdophosphoric blue color procedure (Jackson, 1965). This method is described by Watanabe and Olsen (1965).

3.2. Arsenate present in the extract will produce a blue color with the molybdophosphoric blue color procedure unless the arsenate is reduced. Jackson (1965) describes a reduction procedure.

4. Precision and Accuracy

4.1. Repeated analyses of two standard soil samples over 30 days in the Georgia Soil Testing and Plant Analysis Laboratory gave coefficients of variation of 6.4% and 9.0%, respectively. The mean value for each soil was 32 and 40 kg ha^{-1}, respectively. The variance is essentially a factor related to the heterogeneity of the soil rather than to the extraction or UV-VIS spectrophotometric procedures.

5. Apparatus

5.1. No. 10 (2-mm opening) sieve.
5.2. 4-cm^3 scoop, volumetric.
5.3. 50-mL extraction bottle or flask, with stopper.
5.4. Mechanical reciprocating shaker, 180 oscillations per minute.
5.5. Filter funnel, 11 cm.
5.6. Funnel rack.
5.7. Whatman No. 1 filter paper (or equivalent), 11 cm.
5.8. UV-VIS spectrophotometer set 880 nm.
5.9. Spectrophotometric tube or cuvet.
5.10. Analytical balance.
5.11. Volumetric flasks and pipettes as required for preparation of reagents, standard solutions, and color development.

6. Reagents (use reagent grade chemicals and pure water)

6.1. *Extraction Reagent* (0.05 N HCl in 0.025 N H_2SO_4): Pipette 4.3 mL conc. hydrochloric acid (HCl) and 0.7 mL conc. sulfuric acid (H_2SO_4) into a 1,000 mL volumetric flask and bring to volume with water.

6.2. *Ascorbic Acid Solution:* Weigh 176.0 g ascorbic acid into a 2,000-mL volumetric flask and bring to volume with water. Store in a dark glass bottle in a refrigerated compartment.

6.3. *Sulfuric-Molybdate-Tartrate Solution:* Weigh 100 g ammonium molybdate $[(NH_4)Mo_7O_{24} \cdot 4H_2O]$ into 500 mL water in a 2,000-mL volumetric flask, dissolve, and weigh in 2.425 g antimony potassium tartrate $[K(SbO)C_4H_4O_6 \cdot 1/2H_2O]$. Add slowly 1,400 mL conc. sulfuric acid (H_2SO_4) and mix well. Let it cool and bring to volume with water. Store in a polyethylene or pyrex bottle in a dark, refrigerated compartment.

6.4. *Working Solution:* Add 500 mL water into a 1,000-mL volumetric flask, pipette 10 mL Ascorbic Acid Solution (see 6.2) and 20 mL Sulfuric-Molybdate-Tartrate Solution (see 6.3) into the 1,000-mL volumetric flask and bring to volume with water. Allow it to stand at least 1 hour before using. Store in refrigerator in opaque plastic bottle. The solution is stable for two to three days.

6.5. *Phosphorus Standard* (1,000 mg P L^{-1}): Commercially prepared standard, or weigh 3.85 g ammonium dihydrogen phosphate ($NH_4H_2PO_4$) into a 1,000-mL volumetric flask and bring to volume with Extraction Reagent (see 6.1). Prepare standards containing 1, 2, 5, 10, 15, and 20 mg P L^{-1} diluting aliquots of the 1,000 mg P L^{-1} standard with Extracting Reagent (see 6.1).

7. Procedure

7.1. *Extraction:* Weigh 5 g or scoop 4-cm^3 (see 5.2) of air-dry <10-mesh (2-mm) soil into a 50-mL extraction bottle (see 5.1), add 25 mL Extraction Reagent (see 6.2), and shake for 5 minutes on a reciprocating shaker at a minimum of 180 oscillations per minute (see 5.2). Filter and collect the extract.

7.2. *Color Development:* Pipette 2 mL extractant (see 7.1) into a spectrophotometer cuvet (see 5.9), add 23 mL Working Solution (see 6.4), mix well, and let it stand for

20 minutes. Read the absorbance at 880 nm with the spectrophotometer. The UV-VIS spectrophotometer should be zeroed against a blank consisting of the Extraction Reagent (see 6.1).

8. Calibration and Standards

8.1. *Working Phosphorus Standards:* Using the Phosphorus Standard (see 6.5), prepare six working standard solutions containing from 1 to 20 mg P L^{-1} in the final volume. Make all dilutions with the Extraction Reagent (see 6.1). Use a 2.0-mL aliquot of each standard and carry through the color development (see 7.2).

8.2. *Calibration Curve:* On semi-log graph paper, plot absorbance on the logarithmic scale versus mg P L^{-1} on the linear scale.

8.3. *Time Factor:* Color intensity reaches a maximum in approximately 20 minutes and will remain constant for about six to eight hours.

9. Calculation

9.1. When the sample is weighed, mg P L^{-1} in filtrate times 10 equals 1 mg P ha^{-1}. When the sample is scooped, the results are reported as kg P ha^{-1} for a 20-cm depth of soil. If the extraction filtrate is diluted, the dilution factor must be applied.

10. Effects of Storage

10.1. Soils may be stored in an air-dry condition for several months with no effect on extractable P. The influence of time and temperature is discussed by Houba and Novozamsky (1998).

10.2. After extraction, the extraction solution containing P should not be stored more than 24 hours.

11. Interpretation

11.1. Accurate fertilizer recommendations for P must be based on known field responses based on local soil-climate-crop conditions (Peck et al., 1977; Brown, 1987; Dahnke and Olson, 1990; Withers and Sharpley, 1995). For most soils and crops, the amount of P extracted is to be interpreted as follows:

	Extractable P in Soil	
Category	kg ha^{-1}	1b A^{-1}
very low	<11	<10
low	11–33	10–30
medium	34–67	31–60
high	68–112	61–100
very high	>112	>100

11.2. Interpretations may vary somewhat, depending on soil characteristics and different crops. Interpretative data is given in Thomas and Peaslee (1973).

6.4 MEHLICH NO. 3 EXTRACTION

1. Principle of the Method

1.1. The extraction of phosphorus (P) by this procedure is designed to be applicable across a wide range of soil properties ranging in reaction from acid to basic. This method correlates well with Bray 1 (Bray and Kurtz, 1954) on acid to neutral soil ($r^2 = 0.966$). It does not correlate with Bray P1 on calcareous soils. The Mehlich No. 3 Mehlich, 1984) method correlates with the Olsen extractant (Olsen et al., 1954) on calcareous soils ($r^2 = 0.918$), even though the quantity of Mehlich No. 3-extractable P is considerably higher. Similar comparisons have been made between the commonly used P extraction procedures by Hanlon and Johnson (1984), Schmisek et al. (1989), Sims (1989), and Wolf and Baker (1989).

1.2. This extractant was developed by Dr. Adolf Mehlich in 1981 and described by A. L. Hatfield for the late Dr. Mehlich (Hatfield, 1972). This procedure, a modification of the Mehlich No. 2 extractant (1978), was developed on a 1:10 soil-solution ratio (2.5 cm^3 soil + 25 cm^3 extractant) for a five-minute shaking period at 200, 4-cm reciprocations per minute. The method is described by Tucker (1992), Anon. (1994a), Wolf and Beegle (1995), and Frank et al. (1998).

2. Range, Sensitivity, and Methods of Analysis

2.1. Phosphorus can be extracted and determined in soil concentrations 2 to 400 kg P ha^{-1} without dilution, using the molybdophosphic blue color procedure first described by Murphy and Riley (1962) and modified by Watanabe and Olsen (1965).

2.2. The commonly used method of analysis is UV-VIS spectrophotometry (Wright and Stuczynski, 1996). Being a multielement extraction procedure, P as well as other elements, can be determined by or inductively coupled plasma emission spectrometry (ICP-AES) (Soltanpour et al., 1996, 1998).

3. Precision and Accuracy

3.1. Repeated analyses of two standard soil samples for 36 separate extractions by the North Carolina Department of Agriculture's Soil Testing Laboratory (Raleigh, NC) gave a variance of 6.28 to 6.39%, respectively. Each soil tested 97 and 132 mg P dm^{-3}, respectively. The variance is most likely related more to the heterogeneity of the samples rather than to measurement, extraction or UV-VIS spectrophotometric procedures.

4. Apparatus

4.1. No. 10 (2-mm opening) sieve.
4.2. 5-cm^3 (volumetric) soil measure and Teflon-coated leveling rod.
4.3. 100-mL extraction bottles, plastic or glass, are suitable.
4.4. Reciprocating shaker (200, 4-cm reciprocations per minute).
4.5. Filter funnels, 11 cm.
4.6. Funnel rack.
4.7. Whatman No. 1 (or equivalent) filter paper, 11 cm.
4.8. Vials, polystyrene plastic, 25 and 50-mL capacity, for collection of extract and color development, respectively.

4.9. Automatic extractant dispenser, 25-mL capacity. Other dispensers or pipettes could be used, depending on preference.
4.10. Dilutor-dispenser apparatus for delivery of sample and color development reagent.
4.11. Volumetric flasks and pipettes are required for preparation of reagents and standard solutions. Pipettes could also be used for color development.
4.12. UV-VIS spectrophotometer set at 880 nm.

5. Reagents (use reagent grade chemicals and pure water)

5.1. *Extraction Reagent:* [0.2 N acetic acid (CH_3COOH); 0.25 N ammonium nitrate (NH_4NO_3); 0.015 N ammonium fluoride (NH_4F); 0.13 N nitric acid (HNO_3); 0.001 M EDTA].

Preparation Procedure (NH_4F-EDTA): Weigh 138.9 g ammonium fluoride (NH_4F) into a 1,000-mL volumetric flask containing 600 mL water, then add 73.05 g EDTA, and bring to volume with water.

Final Extraction Reagent: In a 4,000-mL volumetric flask containing 2,500 mL water, add 80 g ammonium nitrate (NH_4NO_3), add 16 mL NH_4F-EDTA (see above), mix, add 46 mL glacial acetic acid (CH_3OOH) and 3.28 mL conc. nitric acid (HNO_3), and then bring to volume with water. The final pH should be 2.5 ± 0.1.

5.2. *Ascorbic Acid Solution:* Dissolve 176.0 g ascorbic acid ($C_6H_8O_6$) in water and dilute to 2,000 mL with water. Store solution in a dark glass bottle in a refrigerated compartment.

5.3. *Sulfuric-Molybdate-Tartrate Solution:* Dissolve 100 g ammonium molybdate [$(NH_4)_6Mo_7O_{24} \cdot 4H_2O$] in 500 mL water. Dissolve 2.425 g antimony potassium tartrate [$K(SbO)C_4H_4O_6 \cdot 1/2\ H_2O$] in molybdate solution. Add slowly 1,400 mL conc. sulfuric acid (H_2SO_4) and mix well. Let it cool and dilute to 2,000 mL with water. Store the solution in a polyethylene or Pyrex bottle in a dark, refrigerated compartment.

5.4. *Working Solution:* Dilute 10 mL Ascorbic Acid Solution (see 5.2) plus 20 mL of the Sulfuric-Molybdate-Tartrate Solution (see 5.3) with the Extraction Reagent (see 5.1) to make 1,000 mL. Allow the solution to come to room temperature before using. Prepare fresh daily.

5.5. *Phosphorus Standard* (200 mg P L^{-1}): Commercially prepared standard, or dissolve 0.879 g monopotassium phosphate (KH_2PO_4) in approximately 500 mL Extraction Reagent (see 5.1) in a 1,000-mL volumetric flask and bring to volume with Extraction Reagent (see 5.1). Prepare standards containing 1, 2, 5, 10, 15, and 20 mg P L^{-1} by diluting appropriate aliquots of 200 mg P L^{-1} with Extraction Reagent (see 5.1).

6. Procedure

6.1. *Extraction:* Measure 5 cm^3 (see 4.2) of air-dry, 10-mesh (2-mm) soil into a 100-mL extraction bottle (see 4.3), add 50 cm^3 Extraction Reagent (see 5.1), and shake for five minutes on a reciprocating shaker (see 4.4). Filter and collect the extract. For the rationale of using a volume soil measure, refer to Mehlich (1973).

6.2. Color Development: Using a pipette or dilutor dispenser, dilute 1 mL of the sample extract (see 6.1) with 27 mL Working Solution (see 5.4). Allow the color to develop for at least 20 minutes before reading. Read the transmission at 880 nm (2-cm light path probe spectrophotometer) or 882 nm for the standard cuvette-type spectrophotometer.

6.3. The color intensity reaches its maximum in approximately 20 minutes and will remain constant for about six hours.

7. Calibration and Standards

7.1. *Working Phosphorus Standards:* After the calibration curve is established (see 5.5), the instrument can be calibrated for routine analysis using Extraction Reagent (see 5.1) as the blank and the 20 mg P L^{-1} standard (see 5.5). The 20 mg P L^{-1} standard should read 10% T following color development at the 1:27 sample-to-working-solution ratio (see 6.2). The blank solution should be diluted at the same 1:27 ratio as the standards. If the instrument reading is significantly above or below 10% T, check the standard preparation, the sample dispenser, or the dilution ratio between the standard and the working solution. The probe colorimeter technique requires a 1:27 sample-to-working-solution ratio to achieve a 10% T reading at 20 mg P L^{-1}. A linear curve can be obtained by converting %T to optical density (OD). Other types of spectrophotometers may require a different sample-to-working-solution ratio.

8. Calculation

8.1. The results are reported as mg P dm^{-3} (mg P L^{-1} of the standard or P soil extract multiplied by 10). Multiply the mg P dm^{-3} by 2 to obtain the kg P ha^{-1} for a 20-cm depth of soil. If the soil extract requires dilution, multiply the results by the appropriate dilution factor.

8.2. To convert soil test values to other units, see Mehlich (1972, 1974).

9. Interpretation

9.1. Critical P levels proposed by Mehlich (1984) are listed below.

Category	mg P dm^{-3}	kg P ha^{-1}	Expected Crop Response
Very low	<20	<40	definite
Low	20–30	40–60	probable
Medium	31–50	62–100	less likely
High	>50	>100	unlikely

10. Effects of Storage

10.1. After air drying, the extractable P levels in soils remain stable for several months. The influence of storage time and temperature is discussed by Houba and Novozamsky (1998).

Phosphorus 83

10.2. After extraction, the P in the extract should be measured within 72 hours.

6.5 MORGAN EXTRACTION

1. Principle of the Method

1.1. This method is primarily for determining phosphorus (P) content in acid soils with cation exchange capacities of less than 20 milliequivalents per 100 grams.
1.2. This method, first proposed by Morgan (1941), was described in detail by Lunt et al. (1950), and later by Greweling and Peech (1965). The extracting reagent is well buffered at pH 4.8, and, when used in conjunction with activated carbon, yields clear and colorless extracts. The Morgan method was used by several state soil testing laboratories in the northeastern and northwestern United States (Jones, 1973, 1998). The method is described by Wolf and Beegle (1995)

2. Range, Sensitivity, and Methods of Analysis

2.1. Phosphorus can be extracted and determined in soil concentrations from 2 to 100 kg P ha^{-1} without dilution. The upper limit may be extended by diluting the extract prior to spectrohotometric determination.
2.2. The sensitivity varies depending on the method of color development. Greater sensitivity can be obtained with the molybdophosphoric blue color method (Jackson, 1958) as compared to the vanado-molybdophosphoric acid color method (Jackson, 1958). The estimated precision of the method is ±1 mg P kg^{-1}.
2.3. The commonly used method of analysis is UV-VIS spectrophotometry (Watson and Isaac, 1990; Wright and Stuczynski, 1996). Since it is a multielement extraction procedure, P as well as other elements can be determined by inductively coupled plasma emission spectrometry (ICP-AES) (Soltanpour et al., 1996, 1998).

3. Interferences

3.1. With some soils, the extract may be colored, varying from light to dark yellow. If the vanado-molybdophosphoric acid method (Jackson, 1958) is employed, decolorizing is necessary to avoid obtaining high results. Decolorization can be accomplished by including "activated" charcoal in the extraction procedure. Decolorization is not necessary if color development is by the molybdophosphoric blue color procedure (Jackson, 1958; Olsen and Sommers, 1982). A description of the method is given by Watanabe and Olsen (1965).
3.2. Arsenate present in the extractant will produce a blue color with the molybdophosphoric blue color procedure unless the arsenate is reduced. A reduction procedure is given by Jackson (1958).

4. Precision and Accuracy

4.1. Repeated analyses of two standard soil samples over 30 days in the Georgia Soil Testing and Plant Analysis Laboratory gave coefficients of variability of 6.4 to 9.0%, respectively. Each soil tested 32 and 40 kg P ha^{-1}, respectively. The variance factor is essentially related to the homogeneity of the soil rather than to the extraction or spectrophotometric procedures.

5. Apparatus

5.1. No. 10 (2-mm opening) sieve.
5.2. 5-cm^3 scoop, volumetric.
5.3. Extraction bottle or flask, 50-mL with stoppers.
5.4. Mechanical reciprocating shaker, 180 oscillations per minute.
5.5. Filter funnel, 11 cm.
5.6. Whatman No. 1 filter paper (or equivalent), 11 cm.
5.7. UV-VIS spectrophotometer set at 880 nm.
5.8. Spectrophotometric tube or cuvet.
5.9. Funnel rack.
5.10. Analytical balance.
5.11. Volumetric flasks and pipettes as required for the preparation of reagents, standard solutions and color development.

6. Reagents (use reagent grade chemicals and pure water)

6.1. *Extraction Reagent:* Weigh 100 g sodium acetate (NaC$_2$H$_3$O$_2$·3H$_2$O) in about 900 mL water in a 1,000-mL volumetric flask, add 30 mL glacial acetic acid (HC$_2$H$_3$O$_2$), adjust the pH to 4.8, and dilute to volume with water.

6.2. *Ascorbic Acid Solution:* Weigh 176.0 g ascorbic acid in water and dilute to 2,000 mL with water. Store in a dark glass bottle in a refrigerated compartment.

6.3. *Sulfuric-Molybdate-Tartrate Reagent:* Weigh 100 g ammonium molybdate [(NH$_4$)$_6$MO$_7$O$_{24}$·4H$_2$O] in 500 mL of water. Weigh 2.425 g antimony potassium tartrate [K(SbO)C$_4$H$_4$P$_6$·1/2 H$_2$O] into the molybdate solution. Add slowly 1,400 mL conc. sulfuric acid (H$_2$SO$_4$) and mix well. Let cool and dilute to 2,000 mL with water. Store in a polyethylene or Pyrex bottle in a dark refrigerated compartment.

6.4. *Working Solution:* Dilute 10 mL Ascorbic Acid Solution (see 6.2) plus 20 mL Sulfuric-Molybdate-Tartrate Reagent (see 6.3) with Extraction Reagent (see 6.1) to 1,000 mL. Prepare fresh daily. Allow it to stand at least two hours before using.

6.5. *Phosphorus Standard* (100 mg P L^{-1}): Commercially prepared standard, or weigh 0.4394 g monobasic potassium phosphate (KH$_2$PO$_4$) that has been oven dried at 100°C (212°F) into a 1,000 mL volumetric flask. Bring to volume with Extraction Reagent (see 6.1).

7. Procedure

7.1. *Extraction:* Measure 5 cm^3 air-dry <10-mesh (2-mm) soil into a 50 mL Extraction Bottle (see 5.3), add 25 mL Extraction Reagent (see 6.1), and shake for five minutes on a reciprocating shaker at a minimum of 180 oscillations per minute. Filter and collect the extract.

7.2. *Color Development:* Pipette 1 mL extractant into a spectrophotometer cuvet (see 5.8), add 24 mL Working Solution (see 6.4), mix well, and let stand 20 minutes. Read the absorbance at 880 nm. The spectrophotometer should be zeroed against a

Phosphorus

blank consisting of 1 mL Extraction Reagent (see 6.1) and 24 mL Working Solution (see 6.4).

8. Calibration and Standards

8.1. *Working Phosphorus Standards:* Using the Phosphorus Standard (see 6.5), prepare six working standards containing from 1 to 20 mg P L^{-1} in the final volume. Make all dilutions with the Extraction Reagent (see 6.1). Use a 1.0 mL aliquot of each working standard and carry through the color development (see 7.2).
8.2. *Calibration Curve:* On semi-log graph paper, plot the percentage of transmittance on the logarithmic scale versus mg P L^{-1} in the standards on the linear scale.
8.3. *Color Intensity:* Reaches a maximum in approximately 20 minutes and will remain constant for about 24 hours.

9. Calculation

9.1. The results are reported as kg P ha^{-1} for a 20 cm depth of soil: kg P ha^{-1} = mg P L^{-1} in extractant × 10.

10. Effects of Storage

10.1. Soils may be stored in an air-dry condition for several months with no effect on extractable P. The influence of time and temperature is discussed by Houba and Novozamsky (1998).
10.2. After extraction, the extraction solution containing P should not be stored for more than 24 hours.

11. Interpretation

11.1. Accurate fertilizer recommendations for P must be based on field response data conducted under local soil-climate-crop conditions (Peck et al., 1977; Brown, 1987; Dahnke and Olson, 1990; Withers and Sharpley, 1995).
11.2. Interpretations may vary somewhat, depending on soil characteristics and different crops.

6.6 AMMONIUM BICARBONATE–DTPA EXTRACTION

1. Principle of the Method

1.1. The Extraction Reagent is 1 M ammonium bicarbonate (NH$_4$HCO$_3$) in 0.005 M DTPA adjusted to a pH of 7.6 (Soltanpour and Schwab, 1977; Soltanpour and Workman, 1979; Soltanpour, 1991). Upon shaking, the pH rises due to the evolution of carbon dioxide (CO$_2$). As the pH rises, a fraction of the bicarbonate (HCO$_3$) changes to carbonate (CO$_3$). The CO$_3^{2-}$ ions precipitate calcium (Ca) from labile calcium phosphates, thus dissolving labile phosphorus (P) in the 15 minutes of shaking.
1.2. This method is highly correlated with the sodium bicarbonate (NaHCO$_3$) method for P (Olsen et al., 1954). The method is described by Anon. (1994d).

2. Range and Sensitivity

2.1. The range and sensitivity are the same as those for Olsen's $NaHCO_3$ P (Olsen et al., 1954).

2.2. The commonly used method of analysis is UV-VIS spectrophotometry (Watson and Isaac, 1990; Wright and Stuczynski, 1996). Being a multielement extraction procedure, P as well as other elements can be determined by inductively coupled plasma emission spectrometry (ICP-AES) (Soltanpour, 1991; Soltanpour et al., 1998).

3. Interferences

3.1. The Extracting Reagent is unstable with regard to pH and should be kept under mineral oil to prevent a pH change. Ideally, the extraction reagent is best made as needed.

3.2. If the blue phosphomolybdo solution developed from AB-DTPA soil extract has more than 0.3 mg P L^{-1} (30 mg P kg^{-1} in soil) (see 7.3), dilute the soil extract with Extraction Reagent (see 6.1) before taking an aliquot to keep the standard regression equation linear (see 12.2 under comments).

4. Precision and Accuracy

4.1. A coefficient of variability ranging from 5 to 10% can be expected for different determinations. Accuracy is comparable to sodium bicarbonate test for P.

5. Apparatus

5.1. No. 10 (2-mm opening) sieve constructed from stainless steel or Nalgene.
5.2. Analytical balance.
5.3. 125-mL polyethylene conical flasks.
5.4. Eberbach reciprocating shaker (or equivalent), 180 oscillations per minute.
5.5. Whatman No. 42 filter paper (or equivalent), 12.5 cm.
5.6. UV-VIS spectrophotometer set at 880 nm.
5.7. 2.5-cm matching spectrophotometric tubes.
5.8. pH meter readable to 0.01 pH units.
5.9. Funnel rack.
5.10. Accurate automatic diluter.
5.11. Volumetric flasks and pipettes as required for preparation of reagents, and standard solutions.

6. Reagents (use reagent grade chemicals and pure water)

6.1. *Extraction Reagent* [Ammonium bicarbonate (AB)-DTPA]: 0.005 M DTPA (diethylenetriaminepentaacetic acid) solution is obtained by adding 9.85 g DTPA (acid form) to 4,500 mL water in a 5,000-mL volumetric flask. Shake for five hours constantly to dissolve the DTPA. Bring to 5,000 mL with water. This solution is stable with regard to pH. To 900 mL of the 0.005 M DTPA solution, add 79.06 g ammonium bicarbonate (NH_4HCO_3) gradually and stir gently with a rod to facilitate dissolution and to prevent effervescence when bicarbonate is added. The solution is diluted to 1,000 mL with the 0.005 M DTPA solution and mixed gently with a rod. The pH is adjusted to 7.6 using 2 M hydrochloric acid (HCl) with slow agitation with a rod. The

AB-DTPA solution must be stored under mineral oil. Check the pH after storage and adjust it with a 2 M (HCl) dropwise, if necessary. The cumulative volume of HCl added should not exceed 1 mL L^{-1} limit, after which a fresh solution should be prepared.

6.2. Mixed Reagent: Weigh 12.0 g ammonium molybdate [$(NH_4)_6Mo_7O_{24} \cdot 4H_2O$] into 250 mL water. Weigh 0.291 g antimony potassium tartrate [$K(SbO)C_4H_4O_6 \cdot 1/2 H_2O$] into 1,000 mL 5.76 N sulfuric acid (H_2SO_4) [160 mL conc. H_2SO_4 L^{-1}]. Mix the two solutions together thoroughly and make to a 2,000 mL volume with water. Store in a Pyrex bottle painted black or wrapped in a black paper or equivalent and store in a cool place. The mixed reagent will be stable for four months.

6.3. Color Developing Reagent (CDR): Weigh 0.264 g ascorbic acid into 50 mL of the Mixed Reagent (see 6.2). This reagent should be prepared daily to prevent oxidation of ascorbic acid.

7. Procedures

7.1. Extraction Procedure: Weigh 10 g air-dry soil <10-mesh (2-mm) in a 125-mL conical flask, add 20 mL Extracting Reagent (see 6.1), and shake on an Eberbach reciprocal shaker or an equivalent shaker for exactly 15 minutes at 180 cycles per minute with flasks kept open. The extracts are then filtered through Whatman 42 filter paper (Soltanpour and Workman, 1979).

7.2. Phosphorus Determination: Pipette 0.250 mL Phosphorus Standard (see 8.1), soil extract (see 7.1), and AB-DTPA (0.00 P solution, see 6.1) into 2.5-cm matching spectrophotometer tubes. To each tube, add 10 mL water and then 2.25 mL Color Developing Reagent (CDR) (see 6.3). Zero the absorbance on the UV-VIS spectrophotometer at 880 nm with the 0.00 P solution 15 min after adding the CDR. Read the absorbance of the phosphomolybdo blue color in standards and soil extracts 15 to 115 minutes after adding the CDR. The final concentrations of P in the blue solution developed from the standards will be 0.05, 0.10, 0.20, and 0.30 mg P L^{-1} (see 12.2 under comments).

8. Standards

8.1. Phosphorus Standards: Dissolve 1.856 g ammonium dihydrogen phosphate ($NH_4H_2PO_4$) in water into a 500-mL volumetric flask and bring to volume with water to obtain a primary standard of 1,000 mg P L^{-1}. Dilute the primary standard with AB-DTPA (see 6.1) to obtain 2.5, 5.0, 10.0, and 15.0 mg P L^{-1}.

9. Calculations

9.1. mg P kg^{-1} in soil = mg P L^{-1} in extract × 100.

10. Effects of Storage

10.1. Air-drying and storage will not have any significant effect on the levels of nutrients. The influence of time and temperature is discussed by Houba and Novozamsky (1998).

10.2. The Extraction Reagent (see 6.1) must be stored under mineral oil.

11. Interpretation

11.1. The following table gives an interpretation of the index values for P for the AB-DTPA soil test. These are general guidelines and should be verified under different soil-climate-crop-management combinations (Peck et al., 1977; Brown, 1987; Dahnke and Olson, 1990; Withers and Sharpley, 1995).

Index Values for P in Soil	
Category	P (mg kg^{-1})
Low	0–4
Marginal	4–7
Adequate	>7

12. Comments

12.1. The AB-DTPA test (with a 1:2 soil-solution ratio) index values for P given above, regardless of soil texture, approximately account for the reduction in P diffusion coefficient with increasing buffering capacity of soils, while the sodium carbonate (NaHCO$_3$) test of Olsen or the resin test neglects it due to its much narrower soil:solution ratios. To change AB-DTPA values to Olsen test values multiply the former by 1, 2, and 3, for sandy, sandy loam, and loamy sand soils; loam soils; and clay soils; respectively (Soltanpour, 1991).

12.2. We limited the final soil extract blue solution to a P concentration of 0.3 mg L^{-1} to make it stable for 115 minutes. The P limit can be doubled to 0.6 mg L^{-1} (60 mg P kg^{-1} of soil), if the absorbance is read 10 to 15 minutes after adding the CDR (Rodrigues et al., 1994).

6.7 REFERENCES

Anon. 1983a. Determination of phosphorus by Bray P1 extraction, In: *Reference Soil Test Methods for the Southern Region of the United States,* Southern Cooperative Series Bulletin 289, R. A. Isaac (ed.). Athens, GA: University of Georgia College of Agriculture Experiment Stations, 20–24.

Anon. 1983b. Determination of phosphorus by Mehlich I (0.05 N HCl in 0.025 N H$_2$SO$_4$) extraction, In: *Reference Soil Test Methods for the Southern Region of the United States,* Southern Cooperative Series Bulletin 289, R. A. Isaac (ed.), Athens, GA: University of Georgia College of Agriculture Experiment Stations, 15–19.

Anon. 1994a. Extractable phosphorus, potassium, calcium, magnesium, sodium, iron, manganese, and copper: Mehlich 3 method, In: *Plant, Soils, and Water Reference Methods for the Western Region,* Western Regional Extension Publication WREP 125, R. G. Gavlak, D. A. Horneck, and R.O. Miller (eds.), Fairbanks, AK: University of Alaska, 25–26.

Anon. 1994b. Extractable phosphorus: Sodium bicarbonate method (Olsen), In: *Plant, Soils, and Water Reference Methods for the Western Region,* Western Regional Extension

Publication WREP 125, R. G. Gavlak, D. A. Horneck, and R. O. Miller (eds.). Fairbanks, AK: University of Alaska, 21–22.

Anon. 1994c. Extractable phosphorus: Dilute acid-fluoride method (Bray-P1), In: *Plant, Soils, and Water Reference Methods for the Western Region,* Western Regional Extension Publication WREP 125, R. G. Gavlak, D. A. Horneck, and R. O. Miller (eds.). Fairbanks, AK: University of Alaska, 22–24.

Anon. 1994d. Extractable phosphorus, potassium, sulfate-sulfur, zinc, iron, manganese, copper, nitrate (Ammonium bicarbonate-DTPA method, In: *Plant, Soils, and Water Reference Methods for the Western Region,* Western Regional Extension Publication WREP 125, R. G. Gavlak, D. A. Horneck, and R. O. Miller (eds.). Fairbanks, AK: University of Alaska, 24–28.

Anon. 1996. Phosphorus. *Soil Fertility Manual,* Chapter 4. Norcross, GA: Potash and Phosphate Institute. 4:1–14.

Anon. 1998. *Soil Test Levels in North America: Summary Update,* PPI/PPIC/FAR Technical Bulletin 1998–3. Norcross, GA: Potash & Phosphate Institute.

Beegle, D. B. and T. C. Oravec. 1990. Comparison of field calibrations for Mehlich-III P and K with Bray P1 and ammonium acetate K for corn. *Commun. Soil Sci. Plant Anal.* 21:1025–1036.

Black, C. A. 1993. *Soil Fertility Evaluation and Control.* Boca Raton, FL: Lewis Publishers.

Bray, R. H. and L. T. Kurtz. 1945. Determination of total, organic and available forms of phosphorus in soils. *Soil Sci.* 59:39–45.

Brown, J. R. (ed.). 1987. *Soil Testing: Sampling, Correlation, Calibration, and Interpretation,* SSSA Special Publication 21. Madison, WI: Soil Science Society of America.

Dahnke, W. C. and R. A. Olsen. 1990. Soil test correlation, calibration, and recommendation, In: *Soil Testing and Plant Analysis,* 3rd ed., SSSA Book Series No. 3, R. L. Westerman (ed.). Madison, WI: Soil Science Society of America, 45–71.

Fixen, P. E. and J. H. Grove. 1990. Testing soils for phosphorus, In: *Soil Testing and Plant Analysis,* 3rd ed., SSSA Book Series No. 3. R. L. Westerman (ed.), Madison WI: Soil Science Society of America, 141–180.

Frank, K., D. Beegle, and J. Denning. 1998. Phosphorus, In: *Recommended Chemical Soil Test Procedures for the North Central Region,* North Central Regional Research Publication No. 221 (revised), J. R. Brown (ed.). Columbia, MO: Missouri Agricultural Experiment Station SB 1001, University of Missouri, 21–29.

Hanlon, E. A. and G. V. Johnson. 1984. Bray/Kurtz, Mehlich-III, AB/DTPA, and ammonium acetate extractions for P, K, and Mg in four Oklahoma soils. *Commun. Soil Sci. Plant Anal.* 15:277–294.

Hatfield, A. L. 1972. Soil test reporting: A nutrient index system. *Commun. Soil Sci. Plant Anal.* 3:425–436.

Houba, V. J. G. and I. Novozamsky. 1998. Influence of storage time and temperature of air-dried soils on pH and extractable nutrients using 0.01 M $CaCl_2$. *Fresenius J. Anal. Chem.* 360:362–365.

Jackson, M. L. 1958. *Soil Chemical Analysis.* Englewood Cliffs, NJ: Prentice-Hall.

Jones, J. B., Jr. 1973. Soil testing in the United States. *Commun. Soil Sci. Plant Anal.* 4:307–322.

Jones, J. B., Jr. 1998. *Plant Nutrition Manual.* Boca Raton, FL: CRC Press, Inc.

Jones, J. B., Jr. 1998. Soil test methods: Past, present, and future use of soil extractants. *Commun. Soil Sci. Plant Anal.* 29:1543–1552.

Khasawneh, F. E., et al. (eds.). 1980. *The Role of Phosphorus in Agriculture.* Madison, WI: American Society of Agronomy.

Kuo, S. 1996. Phosphorus, in: *Methods of Soil Analysis, Part 3, Chemical Methods,* SSSA Book Series No. 5, R.L. Sparks (ed.). Madison, WI: Soil Science Society of America, 869–919.

Kurtz, L. T. 1942. Elimination of fluoride interference in molybdenum blue reaction. *Ind. Eng. Chem. Anal. Ed.* 14:855.

Lindsay, W. L. and E. C. Moreno. 1960. Phosphate phase equilibria in soils. *Soil Sci. Soc. Amer. Proc.* 24:177–182.

Ludwick, A. E. (ed.). 1998. *Western Fertilizer Handbook.* Second Edition. Danville, IL: Interstate Publishers, Inc.

Lunt, H. A., C. L. W. Swanson, and H. G. M. Jacobson. 1950. *The Morgan Soil Testing System.* Connecticut Agricultural Experiment Station (New Haven) Bulletin 541.

Maynard, D. N. and G. J. Hochmuth. 1997. *Knott's Handbook for Vegetable Growers.* 4th ed. New York: John Wiley & Sons,

Mehlich, A. 1953. *Determination of P, Ca, Mg, K, Na, and NH_4.* Raleigh, NC: Mimeo. North Carolina Soil Test Division.

Mehlich, A. 1972. Uniformity of expressing soil test results on a volume basis. *Commun. Soil Sci. Plant Anal.* 3:417–424.

Mehlich, A. 1974. Uniformity of soil test results as influenced by extractants and soil properties, In: *Proceedings Seventh International Colloquium Plant Analysis and Fertilizer Problems,* J. Wehrmann (ed.). Hanover, Germany, 295–305.

Mehlich, A. 1978. New extractant for soil test evaluation of phosphorus, potassium, magnesium, calcium, sodium, manganese, and zinc. *Commun. Soil Sci. Plant Anal.* 9:477–492.

Mehlich, A. 1984. Mehlich 3 soil test extractant: A modification of Mehlich 2 extractant. *Commun. Soil Sci. Plant Anal.* 15(12):1409–1416.

Moody, P. W. and M. D. A. Bolland. 1999. Phosphorus, In: *Soil Analysis: An Interpretation Manual,* K. I. Peverill, L. A. Sparrow, and D. J. Reuter (eds.). Collingwood, Australia: CSIRO Publishing, 187–200.

Morgan, M. F. 1941. *Chemical Soil Diagnosis by the Universal Soil Testing System.* Connecticuit Agricultural Experiment Station Bulletin 450.

Murphy, J. and J. R. Riley. 1962. A modified single solution method for the determination of phosphate in natural waters. *Anal. Chem. Acta.* 27:31–36.

Nelson, W. L., A. Mehlich, and E. Winters. 1953. The development, evaluation and use of soil tests for phosphorus availability, In: *Soil and Fertilizer Phosphorus,* Agronomy No. 4, W. H. Pierre and A. G. Norman (eds.). Madison, WI: American Society of Agronomy, 153–188.

Olsen, S. R. and L. A. Dean. 1965. Phosphorus, In: *Method of Soil Analysis,* Agronomy No. 9. C. A. Black (ed.). Madison, WI: American Society of Agronomy, 1044–1046.

Olsen, S. R. and L. E. Sommers. 1982. Phosphorus, In: *Methods of Soil Analysis,* Part 2, 2nd ed., A. L. Page et al. (ed.). Madison, WI: American Society of Agronomy, 403–430.

Olsen, S. R., C. V. Cole, F. S. Watanabe, and L. A. Dean. 1954. *Estimation of Available Phosphorus in Soils by Extraction with Sodium Bicarbonate, U.S.D.A. Circular No. 939.* Washington, DC: U.S. Government Printing Office.

Peck, T. R., J. T. Cope, Jr., and D. A. Whitney (eds.). 1977. *Soil Testing: Correlating and Interpreting the Analytical Results,* ASA Special Publication 29. Madison, WI: American Society of Agronomy.

Radojevic, M. and V. N. Bashkin. 1999. *Practical Environmental Analysis.* Cambridge, UK.: The Royal Society of Chemistry.

Rechcigl, J. E., G. G. Payne, and C. A. Sanchez. 1992. Comparison of various soil drying techniques on extractable nutrients. *Commun. Soil Sci. Plant Anal.* 23: 2347–2363.

Rodriguez, J. B., J. R. Self, and P. N. Soltanpour. 1994. Optical conditions for phosphorus analysis by the ascorbic acid-molybdenum blue method. *Soil Sci. Soc. Amer. J.* 58:866–870.

Schmisek, M. E., L. J. Cihacek, and L. J. Swenson. 1998. Relationships between the Mehlich-

III soil test extraction procedure and standard soil test methods in North Dakota. *Commun. Soil Sci. Plant Anal.* 29:1719–1729.
Sims, J. T. 1989. Comparison of Mehlich 1 and Mehlich 3 extractants for P, K, Ca, Mg, Mn, Cu, and Zn in Atlantic coastal plain soils. *Commun. Soil Sci. Plant Anal.* 20:1707–1726.
Smith, F. W., B. G. Ellis, and J. Grava. 1957. Use of acid-fluoride solutions for extraction of available phosphorus in calcareous soils and in soils which rock phosphate has been added. *Soil Sci. Soc. Amer. Proc.* 21:400–404.
Soltanpour, P. N. 1991. Determination of nutrient element availability and elemental toxicity by the AB-DTPA soil test and ICPS. *Adv. Soil Sci.* 16:165–190.
Soltanpour, P. N. and A. P. Schwab. 1977. A new soil test for simultaneous extraction of macro- and micro-nutrients in alkaline soils. *Commun. Soil Sci. Plant Anal.* 8:195–207.
Soltanpour, P. N. and S. M. Workman. 1979. Modification of NH_4HCO_3-DTPA soil test to omit carbon black. *Commun. Soil Sci. Plant Anal.* 10:1411–1420.
Soltanpour, P. N., A. Khan, and A. P. Schwab. 1979. Effect of grinding variables on the NH_4HCO_3-DTPA soil test values for Fe, Zn, Mn, Cu, P, and K. *Commun. Soil Sci. Plant Anal.* 10:903–909.
Soltanpour, P. N., C. W. Johnson, S. M. Workman, J. B. Jones, Jr., and R. O. Miller. 1996. Inductively coupled plasma emission spectrometry and inductively coupled plasma-mass spectroscopy, In: *Methods of Soil Analysis, Part 3—Chemical Methods*, SSSA Book Series No. 5. D. L. Sparks (ed.). Madison, WI: Soil Science Society of America, 91–139.
Soltanpour, P. N., C. W. Johnson, S. M. Workman, J. B. Jones, Jr., and R. O. Miller. 1998. Advances in ICP emission and ICP mass spectrometry. *Adv. Agron.* 64:28–113.
Thomas, G. W. and D. E. Peaslee, 1973. Testing soils for phosphorus, In: *Soil Testing and Plant Analysis*, revised ed., L. M. Walsh and J. D. Beaton (eds.).Madison, WI: Soil Science Society of America, 115–132.
Tucker, M. R. 1992. Determination of phosphorus by Mehlich 3 extraction, In: *Reference Soil and Media Diagnostic Procedures for the Southern Region of the United States*, Southern Cooperative Series Bulletin No. 374, S.J. Donohue (ed.). Blacksburg, VA: Virginia Agricultural Experiment Station, 6–8.
Watanabe, F. S. and S. R. Olsen. 1965. Test of an ascorbic acid method for determining phosphorus in water and $NaHCO_3$ extracts from soil. *Soil Sci. Soc. Amer. Proc.* 29:677–678.
Watson, M. E. and R. A. Isaac. 1990. Analytical instruments for soil and plant analysis, In: *Soil Testing and Plant Analysis*, 3rd ed., SSSA Book Series No. 3. R. L. Westerman (ed.). Madison, WI: Soil Science Society of America, 691–740.
Withers, P. J. and A. N. Sharpley. 1995. Phosphorus fertilizers, In: *Soil Amendments and Environmental Quality*. J. E. Rechcigl (ed.). Boca Raton, FL: Lewis Publishers, 65–107.
Wright, R. J. and T. I. Stuczynski. 1996. Atomic absorption and flame emission spectrometry, In: *Methods of Soil Analysis,* Part 3, Chemical Methods, SSSA Book Series No. 5. D. L. Sparks (ed.). Madison, WI: Soil Science Society of America, 65–90.
Wolf, B. 1982. An improved universal extracting solution and its use for diagnosising soil fertility. *Commun. Soil Sci. Plant Anal.* 13:1005–1033.
Wolf, A. M. and D. E. Baker. 1985. Comparison of soil test phosphorus by Olsen, Bray P1, Mehlich 1 and Mehlich 3 methods. *Commun. Soil Sci. Plant Anal.* 16:467–484.
Wolf, A. M. and Douglas Beegle. 1995. Recommended soil tests for the macronutrients: Phosphorus, potassium, calcium, and magnesium, In: *Recommended Soil Testing Procedures for the Northeastern United States,* Northeast Regional Publication No. 493 (revised), J. T. Sims and A. M. Wolf (eds.). Newark, DE: Agricultural Experiment Station, University of Delaware, 30–38.

7 Major Cations (Potassium, Calcium, Magnesium, and Sodium)

Potassium (K), calcium (Ca), magnesium (Mg), and sodium (Na) primarily exist in the soil as exchangeable cations on the collodial complex of the soil for acid to neutral pH soils. Their extraction from the soil is achieved by the use of an "exchange" cation in the extraction reagent, such as ammonium (NH_4^+), hydrogen (H^+), or sodium (Na^+). The extraction reagent is made of sufficient strength to remove most, if not all, of the cations on the collodial complex. Water and weak salt solutions, such as calcium chloride (0.01 M $CaCl_2$), extract those ions in the soil solution that are in equilibrium with those on the exchange complex.

A summary of the potassium (K) soil test level in North American cropland soils for 1997 has been published by the Potash & Phosphate Institute (1998). The summary data show that the number of soils testing "medium" or less has been declining from around 70–80% in the late 1960s to 40–60% in 1997. The basic concepts of K soil chemistry are given in the Potash & Phosphate Institutes's *Soil Fertility Manual* (Anon., 1996) and plant chemistry concepts by Jones (1998a), while soil testing procedures and their interpretation have been described by Haby et al. (1990), Helmke and Sparks (1996), and Suarez (1996). The major cations are discussed in a new book (Peverill et al., 1999) in which the chapters on K, Ca, and Mg are written by Gourley (1999), Bruce (1999), and Aitken and Scott (1999), respectively. Comparisons among cation extraction methods are given by Hanlon and Johnson (1984), Sims (1989), Beegle and Oravec (1990), and Schmisek et al. (1998). The concept of cation ratio and sufficiency level for soil test interpretation is discussed by McLean (1977) and Dahnke and Olson (1990), and a nutrient index system by Hatfield (1972).

7.1 NEUTRAL NORMAL AMMONIUM ACETATE EXTRACTION

1. Principle of the Method

1.1. This method uses a neutral salt solution to replace the cations present in the soil exchange complex; therefore, the cation concentrations determined by this method are referred to as "exchangeable" for non-calcareous soils. For calcareous soils, the cations are referred to as "exchangeable plus soluble."

1.2. The use of neutral normal ammonium acetate (1 N $NH_4C_2H_3O_2$, pH 7.0) to determine exchangeable potassium (K) was first described by Prianischnikov (1913). Schollenberger and Simon (1945) describe the advantages of this extracting reagent as to its effectiveness in wetting soil, replacing exchangeable cations, ease of volatility during analysis, and suitability for use with flame emission spectrophotometry. More recently, this method was described by Jackson (1958), Chapman (1965), Hanlon and Johnson (1984), Haby et al. (1990), Bates and Richards (1993), Simard (1993), Anon. (1994c), Helmeke and Sparks (1996), and Warncke and Brown (1998). The 1 N $NH_4C_2H_3O_2$, pH 7.0 extraction procedure is the most commonly used extraction reagent for determining K, magnesium (Mg), calcium (Ca), and sodium (Na) in soil testing laboratories in the United States (Jones, 1973, 1998b).

2. Range, Sensitivity, and Analysis Methods

2.1. The range of detection will depend on the particular analytical instrument, a range that can be extended by diluting the extract.

2.2. The sensitivity will vary with the type of analytical instrument used, wavelength selected, and method of excitation.

2.3. The commonly used methods of analysis are flame emission or atomic absorption spectrophotometry (AAS) (Watson and Isaac, 1990; Wright and Stuczynski, 1996), or inductively coupled plasma emission spectrometry (ICP-AES) (Soltanpour et al., 1996, 1998). The use of an AutoAnalyzer has been described by Isaac and Jones (1970, 1971) and Flannery and Markus (1972).

3. Interferences

3.1. Under certain conditions, the Extraction Reagent (see 6.1) will extract more K, Mg, Ca, and/or Na than is found in exchangeable form, such as those elements released by weathering action during the period of extraction, and Ca and Mg released through the dissolution of the carbonate form of these elements. The additional amounts of these elements will not usually or significantly alter an analysis for soil fertility evaluation. If the cation exchange capacity (CEC) is estimated by cation summation and the percentage of base saturation is used to assess the fertility status of the soil, then the interference is significant.

3.2. Known interferences and compensation for the changing characteristics of the extract to be analyzed must be acknowledged. The use of internal standards, such as lithium (Li) and compensating elements such as lanthanum (La), is essential in most flame methods of excitation (Watson and Isaac, 1990; Wright and Stuczynski, 1996).

4. Precision and Accuracy

4.1. Repeated analysis of the same soil with medium concentration ranges of K, Ca, Mg, and Na will produce coefficients of variations of 5%-10%. A major portion of the variance is related to the heterogeneity of the soil rather than to the extraction or method of analysis.

4.2. The level of exchangeable K will increase upon air drying of some soils (Sparks, 1987). Soil samples can be extracted in the moist state. However, the difficulties inherent in handling and storing moist soil hinder the easy adaptation of this method

Major Cations (Potassium, Calcium, Magnesium, and Sodium) 95

to a routine method of analysis. Compensation can be made, based on the expected release of K by the particular soil being tested.

5. Apparatus

5.1. No. 10 (2-mm opening) sieve.
5.2. 4.25-cm^3 scoop, volumetric.
5.3. 50-mL extraction bottle or flask, with stoppers.
5.4. Mechanical reciprocating shaker, 180 oscillations per minute.
5.5. Filter funnel, 11 cm.
5.6. Funnel rack.
5.7. Whatman No. 1 filter paper (or equivalent), 11 cm.
5.8. Flame emission, atomic absorption spectrophotometer (AAS), inductively coupled plasma emission spectrometer (ICP-AES), or AutoAnalyzer system.
5.9. Analytical balance.
5.10. Volumetric flasks and pipettes as required for preparation of reagents and standard solutions.

6. Reagents (use reagent grade chemicals and pure water)

6.1. *Extraction Reagent:* Dilute 57 mL glacial acetic acid ($HC_2H_3O_2$) with water to a volume of approximately 500 mL. Then add 69 mL conc. ammonium hydroxide (NH_4OH). *CAUTION:* Use a fumehood. Add sufficient water to obtain a volume of 990 mL. After thoroughly mixing the solution, adjust the pH to 7.0 using either 3 N ammonium hydroxide (NH_4OH) or 3 N $HC_2H_3O_2$. Dilute to a final volume of 1,000 mL with water.

Alternate Method: Weigh 77.1 g ammonium acetate ($NH_4C_2H_3O_2$) in about 900 mL water in a 1,000-mL volumetric flask. After thoroughly mixing the solution, adjust the pH to 7.0 using either 3 N $HC_2H_3O_2$ or 3 N NH_4OH. Bring to volume with water.

6.2. *Potassium Standard* (1,000 mg K L^{-1}): Commercially prepared standard, or weigh 1.9080 g potassium chloride (KCl) into a 1,000-mL volumetric flask and bring to volume with Extraction Reagent (see 6.1). Prepare the working standards by diluting aliquots of this stock solution standard with Extraction Reagent (see 6.1) to cover the anticipated range in concentration to be found in the soil extraction filtrate. Working standards from 5 to 100 mg K L^{-1} should be sufficient for most soils.

6.3. *Calcium Standard* (1,000 mg Ca L^{-1}): Commercially prepared standard, or weigh 2.498 g calcium carbonate ($CaCO_3$) into a 1,000-mL volumetric flask. Add 50 mL water and add dropwise a minimum volume (approximately 20 mL) conc. hydrochloric acid (HCl) to effect the complete solution of the $CaCO_3$. Dilute to the mark with Extraction Reagent (see 6.1). Prepare the working standards by diluting aliquots of this stock solution standard with Extraction Reagent (see 6.1) to cover the anticipated range in concentration to be found in the soil extraction filtrate and to fit the working range of the instrument.

6.4. *Magnesium Standard* (1,000 mg Mg L^{-1}): Commercially prepared standard, or weigh 1.000 g magnesium (Mg) ribbon into a 1,000-mL volumetric flask and dissolve in a minimum volume of dilute HCl. Dilute to 1,000 mL with Extraction Reagent (see 6.1). Prepare working standards by diluting aliquots of this stock solution standard with Extraction Reagent (see 6.1) to cover the anticipated range in concentration to be found in the soil extraction filtrate and to fit the optimum range of the instrument.

6.5. *Sodium Standard* (1,000 mg Na L^{-1}): Commercially prepared standard, or weigh 2.542 g sodium chloride (NaCl) into a 1,000-mL volumetric flask and bring to volume with Extraction Reagent (see 6.1). Prepare working standards by diluting aliquots of this stock solution standard with Extraction Reagent (see 6.1) to cover the anticipated range in concentration to be found in the soil extraction filtrate. Working standards from 1 to 10 mg Na L^{-1} should be sufficient for most soils.

7. Procedure

7.1. *Extraction:* Weigh 5 g or scoop 4.25 cm^3 (see 5.2) air-dry < 10-mesh (2-mm) soil into a 50-mL extraction bottle (see 5.3), add 25 mL Extraction Reagent (see 6.1), and shake for five minutes on a reciprocating shaker (see 5.4). Filter and collect the filtrate.

7.2. *Analysis:* K, Mg, Ca, and Na in the filtrate can be determined by either flame emission, AAS, ICP-AES, or by use of an AutoAnalyzer (see 2.3).

8. Calibration and Standards

8.1. *Working Standards:* Prepare as described in Section 6. If element concentrations are found outside the range of the instrument or standards, suitable dilutions should be prepared, starting with a 1:1 soil extract-to-Extraction Reagent (see 6.1) dilution. Dilution should be made so as to minimize the magnification of errors introduced by diluting.

8.2. *Calibration:* Procedures vary with instrument techniques and type. Every precaution should be taken to use proper procedures and to follow the manufacturer's recommendations in the operation and calibration of the analytical instrument used.

9. Calculation

9.1. The results are reported as kg ha^{-1} for a 20-cm depth of soil: kg element ha^{-1} = mg L^{-1} of element in extraction filtrate × 10.

9.2. To convert to other units for comparison, see Mehlich (1972, 1974).

10. Effects of Storage

10.1. Soils may be stored in an air-dry condition for several months with no effects on the exchangeable K, Mg, Ca, and Na content. Potassium may be released or fixed upon drying for some soils (Sparks, 1987). The influence of storage time and temperature is discussed by Houba and Novozamsky (1998).

10.2. After extraction, the filtrate containing K, Mg, Ca, and Na should not be stored for longer than 24 hours unless it is refrigerated or treated to prevent bacterial growth.

11. Interpretation

11.1. An evaluation of the analytical results for determination of fertilizer recommendations, particularly for the elements K and Mg, must be based on field response data conducted under local soil-climate-crop conditions (Peck et al., 1977; Brown,

1987; Dahnke and Olson, 1990; Haby et al., 1990; Black, 1993; Mikkelsen and Camberato, 1995; Maynard and Hochmuth, 1997; Ludwick, 1998; Reid, 1998).

7.2 MEHLICH NO. 1 (DOUBLE ACID) EXTRACTION

1. Principle of the Method

1.1. This method is primarily used to determine potassium (K), calcium (Ca), magnesium (Mg), and sodium (Na) in soils which have exchange capacities of less than 10 milliequivalents per 100 g, are acid (pH less than 6.5) in reaction, and are relatively low (less than 5%) in organic matter content. The method is not suited for alkaline soils.

1.2. The use of Mehlich No. 1 as an extraction reagent was first described by Mehlich (1953) and then published specifically as a phosphorus (P) extraction reagent by Nelson et al. (1953) as the North Carolina Double Acid (now Mehlich No. 1) method. It is adaptable to the coastal plain soils of eastern United States. It is currently being used by a number of state soil testing laboratories in the United States (Alabama, Delaware, Florida, Georgia, Maryland, New Jersey, South Carolina, and Virginia) (Jones, 1973, 1998b).

2. Range, Sensitivity, and Analysis Methods

2.1. K, Ca, Mg, and Na can be extracted and determined in soil concentrations (kg ha^{-1}) from 50 to 400 for K, 120 to 1200 Ca, and 40 to 360 Mg without dilution. The range and upper limits may be extended by diluting the extracting filtrate prior to analysis.

2.2. The sensitivity will vary with the type of analytical instrument used, wavelength selected, and method of excitation.

2.3. The commonly used methods of analysis are flame emission or atomic absorption spectrophotometry (AAS) (Watson and Isaac, 1990; Wright and Stuczynski, 1996), or by inductively coupled plasma emission spectrometry (ICP-AES) (Soltanpour et al., 1996, 1998). The use of an AutoAnalyzer has been described by Isaac and Jones (1970, 1971) and Flannery and Markus (1972).

3. Interferences

3.1. Known interferences and compensation for the changing characteristics of the extract to be analyzed must be acknowledged. The use of internal standards, such as lithium (Li) and compensating elements such as lanthanum (La), are essential in most flame methods of excitation (Watson and Isaac, 1990; Wright and Stuczynski, 1996).

4. Precision and Accuracy

4.1. Repeated analysis of the same soil with medium concentration ranges of K, Ca, Mg, and Na will produce coefficients of variation from 5%-10%. A major portion of the variance is related to the heterogeneity of the soil rather than to the extraction or method of analysis.

4.2. The level of exchangeable K will increase upon air drying of some soils (Sparks, 1987). Soil samples can be extracted in the moist state. However, the difficulties inherent in handling and storing moist soil hinder the easy adaptation of this method to a routine method of analysis. Compensation can be made based on the expected release of K by the particular soil being tested.

5. Apparatus

5.1. No. 10 (2-mm opening) sieve.
5.2. 5-cm^3 scoop, volumetric, stainless steel or hard plastic.
5.3. 50-mL extraction bottle or flask, with stopper.
5.4. Mechanical reciprocating shaker, 180 oscillations per minute.
5.5. Filter funnel, 11 cm.
5.6. Funnel rack.
5.7. Whatman No. 1 filter paper (or equivalent), 11 cm.
5.8. Flame emission, atomic absorption spectrophotometer (AAS), inductively coupled plasma emission spectrometer (ICP-AES), or Auto Analyzer System.
5.9. Analytical balance.
5.10. Volumetric flasks and pipettes as required for preparation of reagents and standard solutions.

6. Reagents (use reagent grade chemicals and pure water)

6.1. *Extraction Reagent* (0.05 N HCl in 0.025 N H_2SO_4): Pipette 4.3 mL conc. hydrochloric acid (HCl) and 0.7 mL conc. sulfuric acid (H_2SO_4) into a 1,000-mL volumetric flask and bring to volume with water.

6.2. *Potassium Standard* (1,000 mg K L^{-1}): Commercially prepared standard, or weigh 1.908 g potassium chloride (KCl) into a 1,000-mL volumetric flask and bring to volume with Extraction Reagent (see 6.1). Prepare working standards by diluting aliquots of this stock solution standard with Extraction Reagent (see 6.1) to cover the anticipated range in concentration to be found in the soil extraction filtrate. Working standards from 5 to 50 mg K L^{-1} should be sufficient for most soils.

6.3. *Calcium Standard* (1,000 mg Ca L^{-1}): Commercially prepared standard, or weigh 2.498 g calcium carbonate ($CaCO_3$) into a 1,000-mL volumetric flask, add 50 mL water, and add dropwise a minimum volume (approximately 20 mL) conc. HCl to effect the complete solution of the $CaCO_3$. Dilute to the mark with Extraction Reagent (see 6.1). Prepare working standards by diluting aliquots of this stock solution standard with Extraction Reagent (see 6.1) to cover the anticipated range in concentration to be found in the soil extraction filtrate. Working standards from 15 to 150 mg Ca L^{-1} should be sufficient for most soils.

6.4. *Magnesium Standard* (1,000 mg Mg L^{-1}): Commercially prepared standard, or weigh 1.000 g magnesium (Mg) ribbon into a 1,000-mL volumetric flask and dissolve in the minimum volume of dilute HCl and dilute to the mark with Extraction Reagent (see 6.1). Prepare working standards by diluting aliquots of this stock solution standard with Extraction Reagent (see 6.1) to cover the anticipated range in concentration to be found in the soil extraction filtrate. Working standards from 5 to 50 mg Mg L^{-1} should be sufficient for most soils.

6.5. *Sodium Standard* (1,000 mg Na L^{-1}): Commercially prepared standard, or weigh 2.542 g sodium chloride (NaCl) into a 1,000-mL volumetric flask and bring to volume with Extraction Reagent (see 6.1). Prepare working standards by diluting aliquots of this stock solution standard with Extraction Reagent (see 6.1) to cover the anticipated range in concentration to be found in the soil extraction filtrate. Working standards from 1 to 10 mg Na L^{-1} should be sufficient.

7. Procedure

7.1. *Extraction:* Weigh 5 g or scoop 5-cm^3 (see 5.2) air-dry, < 10-mesh (2-mm) soil into a 50-mL extraction bottle (see 5.3), add 25 mL Extraction Reagent (see 6.1), and shake for 5 minutes on a reciprocating shaker (see 5.4). Filter and collect the filtrate.

7.2. *Analysis:* K, Ca, Mg, and Na in the filtrate can be determined by either flame emission, AAS, ICP-AES, or by the use of an Auto Analyzer (see 2.3).

8. Calibration and Standards

8.1. *Working Standards:* Prepare as described in Section 6. If element concentrations are found outside the range of the instrument or standards, suitable dilutions should be prepared starting with a 1:1 soil extract to Extraction Reagent (see 6.1) dilution. Dilution should be made so as to minimize the magnification of errors introduced by diluting.

8.2. *Calibration:* Procedures vary with instrument techniques and the type of analytical instrument. Every precaution should be taken to ensure that the proper procedures are followed and that the manufacturer's recommendations are followed in the operation and calibration of the instrument.

9. Calculation

9.1. The results are reported as kg element ha^{-1} for a 20-cm depth of soil: kg element ha^{-1} = mg element L^{-1} in extraction filtrate × 10. If the extraction filtrate is diluted, the dilution factor must be applied.

9.2. To convert to other units for comparisons, see Mehlich (1972, 1974).

10. Effects of Storage

10.1. Soils may be stored in an air-dry condition for several months with no effect on the exchangeable K, Ca, Mg, and Na content. The influence of storage time and temperature is discussed by Houba and Novozamsky (1998).

10.2. After extraction, the filtrate containing K, Ca, Mg, and Na should not be stored longer than 24 hours unless it is refrigerated or treated to prevent bacterial growth.

11. Interpretation

11.1. An evaluation of the analytical results for the determination of fertilizer recommendations, particularly for the elements K and Mg, must be based on field response data conducted under local soil-climate-crop conditions (Peck et al., 1977; Brown, 1987; Dahnke and Olson, 1990; Haby et al. 1990; Black, 1993; Mikkelsen and Camberato, 1995; Maynard and Hochmuth, 1997; Ludwick, 1998; Reid, 1998).

7.3 MEHLICH NO. 3 EXTRACTION

1. Principle of the Method

1.1. The extraction of calcium (Ca), magnesium (Mg), potassium (K), and sodium (Na) by this method is designed to be applicable across a wide range of soil properties ranging in reaction from acid to basic (Tucker, 1992; Sen Tram and Simard, 1993; Warnke and Brown, 1998) and is a modification of Mehlich No. 2 (Mehlich, 1978). The Mehlich No. 3 method correlates well with Mehlich No. 1, Mehlich No. 2, and ammonium acetate (Hanlon and Johnson, 1984; Mehlich, 1984; Schmisek et al., 1989; Sims, 1989; Beegle and Oravec, 1990). For specific extraction values and correlation coefficients, see Mehlich (1978, 1984).

1.2. This extractant was developed by Dr. Adolf Mehlich in 1980–81, and was described by Hatfield for Mehlich (1984). This procedure was developed on a 1:10 soil-solution ratio (2.5 cm^3 soil + 25 cm^3 extractant) for a five-minute shaking period at 200, 4-cm reciprocations per minute. The method is described by Mehlich (1984), Tucker (1992), Anon. (1994b), and Wolf and Beegle (1995).

2. Range and Sensitivity

2.1. Following a 1:4 dilution of the soil extract with the Lithium Working Solution (see 6.2), soil concentrations of K and Na can be determined up to 1,564 and 920 kg ha^{-1}, respectively. Following a 1:10 dilution of the soil extract with the Lanthanum Solution (see 6.3), soil concentrations of Ca and Mg can be determined up to 10,020 and 1,216 kg ha^{-1}, respectively. An atomic absorption spectrophotometer equipped with a 3-standard microprocessor is required to obtain linearity at instrument readings above 100. In the absence of microprocessor-equipped instrumentation, extractable Ca and Mg can be determined in soil concentrations up to 10 and 2 meq 100 cm^{-3}, respectively.

2.2. Sensitivity will vary with the type of instrument, wavelength selection, and method of excitation.

2.3. The commonly used methods of analysis are flame emission, atomic absorption spectrophotometry (Watson and Isaac, 1990; Wright and Stuczynski, 1996), or inductively coupled plasma emssion spectrometry (ICP-AES) (Soltanpour et al., 1996, 1998).

3. Interferences

3.1. Chemical interferences and compensation for changes in the characteristics of the extract to be analyzed must be acknowledged. Internal standards, such as lithium

Major Cations (Potassium, Calcium, Magnesium, and Sodium)

(Li), and compensating elements, such as lanthanium (La), are required for most flame methods of excitation (Watson and Isaac, 1990).

4. Precision and Accuracy

4.1. Repeated analyses of the same soil with medium ranges of K, Ca, and Mg will give variances from 5% to 10%. The major portion of the variance is related more to the heterogeneity of the soil than to measurement, extraction, or method of analysis.

5. Apparatus

5.1. No. 10 (2-mm) sieve.
5.2. 2.5-cm^3 scoop (volumetric) and Teflon-coated leveling rod.
5.3. 100-mL extraction bottles, plastic, or glass.
5.4. Reciprocating shaker (200, 4-cm reciprocations per minute).
5.5. Filter funnels, 11 cm.
5.6. Funnel rack.
5.7. Whatman No. 1 (or equivalent) filter paper, 11 cm.
5.8. Vials, polystyrene plastic, 25-mL capacity, for collection of extract and sample dilutions.
5.9. Automatic dispenser for extractant, 25-mL capacity.
5.10. Diluter-dispenser mechanism or pipettes, 10-mL capacity.
5.11. Flame emission, atomic absorption spectrophotometer (AAS), or inductively coupled plasma emission spectrometer (ICP-AES).
5.12. Volumetric flasks and pipettes as required for preparation of reagents in standard solutions.
5.13. Analytical balance.

6. Reagents (use reagent grade chemicals and pure water)

6.1. *Extraction Reagent* [0.2 N acetic acid (CH$_3$COOH); 0.25 N ammonium nitrate (NH$_4$NO$_3$); 0.015 N ammonium fluoride (NH$_4$F); 0.13 N nitric acid (HNO$_3$); 0.001 M EDTA]:
   Preparation Procedure (NH$_4$F-EDTA): Weigh 138.9 g ammonium fluoride (NH$_4$F) into a 1,000-mL volumetric flask containing 600 mL water, then add 73.05 g EDTA, and bring to volume with water.
   Final Extraction Reagent: In a 4,000-mL volumetric flask containing 2,500 mL water, add 80 g ammonium nitrate (NH$_4$NO$_3$), add 16 mL NH$_4$F-EDTA (see above), mix, add 46 mL glacial acetic acid (CH$_3$OOH) and 3.28 mL conc. nitric acid (HNO$_3$), and then bring to volume with water. The final pH should be 2.5 ± 0.1.
6.2. *Lithium Working Solution:* Dilute 12.5 mL commercial Lithium (Li) Standard (1,500 mg Li L^{-1}) to 1,000 mL with water. This solution is used as an internal standard for the determination of K.
6.3. *Lanthanum Compensation Solution:* Suspend 13 g lanthanum oxide (La$_2$O$_3$) in 50 mL water in a large beaker and dissolve with 28 mL conc. HNO$_3$. Allow the solution to cool, and then pour it into a 2,000-mL volumetric flask and bring to volume with water.

6.4. *Potassium and Sodium Standard:* Commercially prepared standard, or weigh 0.7456 g potassium chloride (KCl) and 0.5844 g sodium chloride (NaCl) in a 1,000-mL volumetric flask and bring to volume with Extraction Reagent (see 6.1). Alternatively, dilute 100 mL commercial K and Na Standard (100 mg K and 100 mg Na^{-1}) to 1,000 mL with Extraction Reagent (see 6.1). Prepare working standards to contain 0, 0.5, 1.0, and 2.0 mg K and Na L^{-1} by appropriate dilution with Extraction Reagent (see 6.1)

6.5. *Calcium and Magnesium Standard:* Commercially prepared standard, or weigh 2.5 g calcium carbonate ($CaCO_3$) and 10.14 g magnesium sulfate ($MgSO_4 \cdot 7H_2O$) in approximately 500 mL Extraction Reagent (see 6.1) and 10 mL conc. HNO_3 and bring to 1,000-mL volume with Extraction Reagent (see 6.1). Alternatively, dilute 500 mL commercial Ca Reference Standard (1 mL = 1 mg Ca) and 60.75 mL commercial Mg Reference Standard (1 mL = 1 mg Mg) to 1,000 mL with Extraction Reagent (see 6.1). Dilute with Extraction Reagent (see 6.1) to obtain 5, 10, 15, 20, and 25 meq Ca, and 1, 2, 3, and 5 meq Mg L^{-1}, respectively. Before making the 5, 10, and 15 meq standards to volume, add 3.0 mL concentrated HNO_3 to prevent Ca precipitation. The latter dilutions compose the working standards.

7. Procedure

7.1. *Extraction:* Measure 2.5 cm^3 air-dry, <10-mesh (2-mm) soil into a 100-mL Extraction Bottle (see 5.3), add 25 cm^3 Extracting Solution (see 6.1), and shake for 5 minutes on a reciprocating shaker (see 5.4). Filter and collect the extract in 25-cm^3 plastic vials (see 5.8).

8. Analysis and Calibration

8.1. *Determination of Potassium and Sodium:* Using a diluter-dispenser (see 5.10) or pipette, transfer 2 mL soil extract (see 7.1) or Potassium-Sodium Working Standards (see 6.4) and 8 mL Lithium Working Solution (see 6.2) into plastic vials (see 5.8). Set the instrument reading at 100 using the 1-mg Potassium-Sodium L^{-1} Working Standard. Atomize soil extract and record the instrument reading. For final calculations for a 20-cm depth of soil to meq 100 cm^{-3}, mg dm^{-3}, kg ha^{-1} and lbs A^{-1} of K and Na (see 7.4).

8.2. *Instrument Calibration:* Proper precautions should be taken to follow the manufacturer's recommendations in the operation and calibration of the instrument. Linearity between the concentration of K and Na can be ascertained by running a series of appropriate standards (see 6.4). If the instrument reading exceeds 200, dilute equal portions of the soil extract-lithium sample mixture with zero standard (see 7.2) and multiply the results by 2.

8.3. *Calculations (Potassium and Sodium):* With the Potassium-Sodium Standard (see 6.4), set the instrument at 100. Multiply the instrument reading (IR) by 0.01 meq K or Na 100 cm^{-3} soil. Alternatively, the IR × 3.91 = mg K dm^{-3} and IR × 2.3 = mg Na dm^{-3} of soil. To convert mg K and Na dm^{-3} to kg ha^{-1}, multiply by 2. Multiply mg K or Na dm^{-3} by 1.78 to obtain lbs A^{-1} to a depth of 20 cm. All of the above calculations are based on a volume of soil (see 5.2) which is employed in these soil test procedures.

8.4. *Determination of Calcium and Magnesium:* Using a diluter-dispenser (see 5.10) or pipette, transfer 1 mL Calcium-Magnesium Standard (see 6.5) or soil extract (see 7.1) and 9 mL Lanthanum Compensating Solution (see 6.3) into plastic vials (see 5.8). Adjust the instrument to zero with a blank composed of 1 mL extractant (see 6.4) and 9 mL Lanthanum Compensating Solution (see 6.6). Standardize the instrument with Ca-Mg Standards (see 6.5) by setting the 10 meq Ca-2 meq Mg Working Standard (see 6.5) at 100 and the 25 meq Ca–5 meq Mg Working Standard at 250 instrument readings, respectively. An instrument equipped with a 3-standard microprocessor is required in order to obtain linearity above an instrument reading of 100.

8.5. *Instrument Calibration:* Proper precautions should be taken to follow the manufacturer's recommendations in the operation and calibration of the instrument. Linearity between the concentration of Ca and Mg can be ascertained by running a series of appropriate standards (see 6.5). Calcium and Mg concentrations are linear up to 10 meq Ca and 2 meq Mg at a corresponding instrument reading of 100. By the use of a 3-standard microprocessor, linearity can be obtained up to 25 meq Ca and 5 meq Mg with a corresponding instrument reading of 250. If scale expansion above the 100 instrument reading is not available and the unknown readings exceed 100, dilute the known aliquot of soil extract-lanthanum mixture (see 6.3) with zero standard and multiply by the dilution factor.

8.6. *Calculations (Calcium and Magnesium):* With the 10 meq Ca-2 meq Mg and 25 meq Ca-5 meq Mg standards (see 6.8), set at an instrument reading (JR) of 100 and 200, respectively. The corresponding concentrations on a volume soil basis are: JR $\times$ 0.1 = meq Ca 100 cm^{-3} and JR $\times$ 0.02 = meq Mg 100 cm^{-3}. Alternatively, to convert JR to mg Ca and Mg dm^{-3}, multiply by 20.04 and 2.432, respectively. To obtain kg Ca and Mg ha^{-1}, multiply mg dm^{-3} by 2. Then kg ha^{-1} of Ca and Mg multiplied by 0.891 = lbs A^{-1}.

9. Calculations

9.1. When 5-cm^3 soil is extracted with 25 mL Extraction Reagent, mg K, Mg, Ca, Na dm^{-3} in the filtrate times 5 equals mg K, Mg, Ca, Na dm^{-3} in soil, then mg dm^{-3} times 2 equals kg K, Mg, Ca, Na ha^{-1} to a depth of 20 cm (7.87 inches). One hectare (ha) contains 2 million cubic decimeters to a depth of 20 cm (7.87 inches). Multiply kg K, Mg, Ca, Na ha^{-1} by 0.894 to obtain lbs K, Mg, Ca, Na A^{-1}. If the extractant filtrate requires dilution, the dilution factor must be applied.

9.2. For comparisons of methods of expression, see Mehlich (1972, 1974).

10. Effects of Storage

10.1. Soils may be stored in an air-dry condition for several months with no effect on exchangeable K, Mg, Ca, and Na content. Potassium may be released or fixed upon drying for some soils (Sparks, 1987). The influence of storage time and temperature is discussed by Houba and Novozamsky (1998).

10.2. After extraction, the filtrate containing K, Mg, Ca, and Na should not be stored longer than 24 hours unless it is refrigerated or treated to prevent bacterial growth.

11. Interpretation

11.1. Evaluation of the analytical results and the corresponding fertilizer recommendations must be based on field response data conducted under local soil-climate crop conditions (Peck et al., 1977; Brown, 1987; Dahnke and Olson, 1990; Haby et al., 1990). Mehlich proposed critical levels of K and Mg for the Mehlich No. 3 extractant (Mehlich, 1984). Interpretative guidelines for evaluating percentage of cations and base saturation (Mehlich, 1978).

7.4 MORGAN EXTRACTION

1. Principle of the Method

1.1. This method is primarily used for determining potassium (K), calcium (Ca), and magnesium (Mg) in acid soils with cation exchange capacities of less than 20 milliequivalents per 100 g.

1.2. This method, initially proposed by Morgan (1941), was described in detail first by Lunt et al. (1950) and later by Greweling and Peech (1965). The Morgan method was used by a number of soil testing laboratories in the northeastern and northwestern United States (Nelson et al., 1953; Jones, 1973, 1998b). The Extraction Reagent is well buffered at pH 4.8 and, when used in conjunction with *activated* carbon, yields clear and colorless extracts. The concentration of sodium acetate is sufficiently high to effect replacement of about 80% of the exchangeable cations. The method has been modified by Wolf (1982) to include additional elements.

2. Range, Sensitivity, and Analysis Methods

2.1. The range of detection will depend upon the setup of the particular analytical instrument. The range can be extended by diluting the extract prior to analysis.

2.2. The sensitivity will vary with the type of analytical instrument used, wave length selected, and method of excitation.

2.3. The commonly used methods of analysis are flame emission and atomic absorption spectrophotometry (Watson and Isaac, 1990; Wright and Stuczynski, 1996), or by inductively coupled plasma emission spectrometry (ICP-AES) (Soltanpour et al., 1996, 1998).

3. Interferences

3.1. Under certain conditions, the Extraction Reagent (see 6.1) will extract more than those elements which exist in the soil only in an exchangeable form, such as, in the soil solution, those elements released by weathering action during the period of extraction, and the dissolution of carbonates of Ca and Mg if these are present in the soil. However, these contributions will not normally alter significantly the analysis when it is used to assess the fertility status of the soil.

3.2. Known interferences and compensation for the changing characteristics of the extract to be analyzed must be acknowledged. The use of an internal standard such as lithium (Li) and a compensating element such as lanthanum (La) is essential in most flame methods of excitation (Watson and Isaac, 1990; Wright and Stuczynski, 1996).

4. Precision and Accuracy

4.1. Repeated analysis of the same soil with medium concentration ranges of K, Ca, and Mg will give coefficients of variability of from 5 to 10%. A major portion of the variance is related to the heterogeneity of the soil rather than to the extraction or method of analysis.

4.2. The level of exchangeable K will increase upon the air drying of some soils (Sparks, 1987). Soil samples can be extracted in the moist state. However, the difficulties inherent in handling and storing moist soil make this method difficult for easy adaptation to a routine method of analysis. Compensation can be made based on the expected release of K by the particular soil being tested.

5. Apparatus

5.1. No. 10 (2-mm opening) sieve.
5.2. 5-cm^3 scoop, volumetric.
5.3. Extraction bottle or flask, 50-mL, with stopper.
5.4. Mechanical reciprocating shaker, 180 oscillations per minute.
5.5. Filter funnel, 11 cm.
5.6. Funnel rack.
5.7. Whatman No. 1 filter paper (or equivalent), 11 cm.
5.8. Flame emission, atomic absorption spectrophotometer (AAS), or inductively coupled plasma emission spectrometer (ICP-AES).
5.9. Analytical balance.
5.10. Volumetric flasks and pipettes as required for preparation of reagents and standard solutions.

6. Reagents (use reagent grade chemicals and pure water)

6.1. *Extraction Reagent:* Weigh 100 g sodium acetate (NaC$_2$H$_3$O$_2$·3H$_2$O) into a 1,000-mL volumetric flask and add about 900 mL water. Add 30 mL glacial acetic acid (HC$_2$H$_3$O$_2$), adjust the pH to 4.8, and bring to volume with water.

6.2. *Potassium Standard* (1,000 mg K L^{-1}): Commercially prepared standard, or weigh 1.908 g potassium chloride (KCl) into a 1,000-mL volumetric flask and bring to volume with Extraction Reagent (see 6.1). Prepare working standards by diluting aliquots of this stock solution standard with the Extraction Reagent to cover the anticipated range in concentration to be found in the soil extraction filtrate. Working standards from 5 to 100 mg K L^{-1} should be sufficient for most soils.

6.3. *Calcium Standard* (1,000 mg Ca L^{-1}): Commercially prepared standard, or weigh 2.498 g calcium carbonate (CaCO$_3$) into a 1,000-mL volumetric flask, add 50 mL water, and add dropwise a minimum volume of concentrated hydrochloric acid (HCl) (approximately 20 mL) to effect complete solution of the CaCO$_3$. Dilute to the mark with Extraction Reagent (see 6.1). Prepare working standards by diluting aliquots of this stock solution standard with Extraction Reagent to cover the anticipated range in concentration to be found in the soil extraction filtrate. Working standards from 50 to 220 mg Ca L^{-1} should be sufficient for most soils.

6.4. *Magnesium Standard* (1,000 mg Mg L^{-1}): Commercially prepared standard, or weigh 1.000 g magnesium (Mg) ribbon into a 1,000-mL volumetric flask and dissolve in a minimum volume of (1+1) HCl and bring to volume with Extraction Reagent (see 6.1). Prepare working standards by diluting aliquots of this stock solution standard with Extraction Reagent to cover the anticipated range in concentration to be found in the soil extraction filtrate. Working standards from 5 to 50 mg Mg L^{-1} should be sufficient for most soils.

7. Procedure

7.1. *Extraction:* Measure 5 cm^3 (see 5.2) of air-dry, < 10-mesh (2-mm) soil into a 50-mL extraction bottle (see 5.3), add 25 mL Extraction Reagent (see 6.1), and shake for five minutes on a reciprocating shaker at a minimum of 180 oscillations per minute (see 5.4). Filter and collect the filtrate.

7.2. *Analysis:* K, Ca, and Mg in the filtrate can be determined by either flame emission, AAS, or ICP-AES (see 2.3).

8. Calibration and Standards

8.1. *Working Standards:* Prepared as described in Section 6. If element concentrations are found outside the range of the analytical instrument or standards, suitable dilutions should be prepared starting with a 1:1 soil extract to Extraction Reagent (see 6.1) dilution. Dilution should be made so as to minimize the magnification of errors introduced by diluting.

8.2. *Calibration Curve:* Procedures vary with instrumental techniques and type of instrument. Every precaution should be taken to ensure that the analytical instrument is properly calibrated and used.

9. Calculation

9.1. The results are reported as kg ha^{-1} for a 20 cm depth of soil: kg of element ha^{-1} = mg element L^{-1} in extraction filtrate × 10.

9.2. To convert to other units for comparison, see Mehlich (1972, 1974).

10. Effects of Storage

10.1. Soils may be stored in an air-dry condition for several months with no effect on the exchangeable K, Ca, and Mg content. Potassium may be released on drying for some soils (Sparks, 1987). The influence of storage time and temperature is discussed by Houba and Novozamsky (1998).

10.2. After extraction, the filtrate containing K, Ca, and Mg should not be stored longer than 24 hours unless it is refrigerated or treated to prevent bacterial growth.

11. Interpretation

11.1. An evaluation of the analysis results as well as accurate fertilizer recommendations, particularly for the elements K and Mg, must be based on field response data conducted under local soil-climate-crop conditions (Peck et al., 1977; Brown, 1987; Dahnke and Olson, 1990; Haby et al., 1990; Black, 1993; Mikkelsen and Camberato, 1995; Maynard and Hochmuth, 1997; Ludwick, 1998; Reid, 1998).

7.5 AMMONIUM BICARBONATE–DTPA EXTRACTION (POTASSIUM)

1. Principle of the Method

1.1. The extraction reagent is 1 M ammonium bicarbonate (NH_4HCO_3) in 0.005 M DTPA adjusted to a pH of 7.6 (Soltanpour and Schwab, 1977; Soltanpour and Workman, 1979; Soltanpour, 1991). The ammonium (NH_4^+) ion will exchange with the potassium (K^+) ion the collodial complex bringing it into solution. The method is described by Anon. (1994a).

1.2. This method is highly correlated with the neutral normal ammonium acetate (1 N $NH_4C_2H_3O_2$, pH 7.0) method for K.

2. Range, Sensitivity, and Analysis Method

2.1. The range and sensitivity are the same as 1 N $NH_4C_2H_3O_2$, pH 7.0 for K.
2.2. The sensitivity will vary with the type of analytical instrument used, wavelength selected, and method of excitation.
2.3. The commonly used methods of analysis are flame emission, atomic absorption spectrophotometry (AAS) (Watson and Isaac, 1990; Wright and Stuczynski, 1996), or inductively coupled plasma emission spectrometry (ICP-AES) (Soltanpour, 1991; Soltanpour et al., 1996, 1998).

3. Interferences

3.1. The extraction reagent is unstable with regard to pH and should be kept under mineral oil to prevent a pH change. Ideally, the extraction reagent should be made as needed.
3.2. Grinding force and time and the amount of soil in the grinder should be adjusted so that extracted levels of trace elements are equivalent to those obtained with a wooden roller (Soltanpour et al., 1979).

4. Precision and Accuracy

4.1. A coefficient of variability ranging from 5 to 10% can be expected for different determinations. Accuracy is comparable to 1 N $NH_4C_2H_3O_2$, pH 7.0 test for K.

5. Apparatus

5.1. No. 10 (2-mm opening) sieve constructed from stainless steel or Nalgene.
5.2. Analytical balance.
5.3. 125-mL polyethylene conical flasks.
5.4. Eberbach reciprocating shaker (or equivalent), 180 oscillations per minute.
5.5. Whatman No. 42 filter paper (or equivalent), 12.5 cm.
5.6. Funnels and funnel rack.

5.7. pH meter readable to 0.01 pH units.

5.8. Flame emission or atomic absorption spectrophotometer (AAS) or inductively coupled plasma atomic emission spectrometer (ICP-AES).

5.9. Accurate automatic diluter.

5.10. Volumetric flasks and pipettes as required for preparation of reagents, and standard solutions.

6. Reagents (use reagent grade chemicals and pure water)

6.1. *Extraction Reagent* [Ammonium bicarbonate (AB)-DTPA]: 0.005 M DTPA (diethylenetriaminepentaacetic acid) solution is obtained by adding 9.85 g DTPA (acid form) to 4,500 mL water in a 5,000-mL volumetric flask. Shake for five hours constantly to dissolve the DTPA. Bring to 5,000 mL with water. This solution is stable with regard to pH. To 900 mL of the 0.005 M DTPA solution, add 79.06 g ammonium bicarbonate (NH_4HCO_3) gradually and stir gently with a rod to facilitate dissolution and to prevent effervescence when bicarbonate is added. The solution is diluted to 1,000 mL with the 0.005 M DTPA solution and mixed gently with a rod. The pH is adjusted to 7.6 using 2 M hydrochloric acid (HCl) with slow agitation with a rod. The AB-DTPA solution must be stored under mineral oil. Check the pH after storage and adjust it with a 2 M HCl solution dropwise, if necessary. The cumulative volume of HCl added should not exceed 1 mL L^{-1} limit, after which a fresh solution should be prepared.

7. Procedure

7.1. *Extraction Method:* Weigh 10 g air-dry soil <10-mesh(2-mm) into a 125-mL conical flask, add 20 mL Extraction Reagent (see 6.1), and shake on an Eberbach reciprocal shaker or an equivalent shaker for exactly 15 minutes at 180 cycles per minute with flasks kept open. The extracts are then filtered through Whatman 42 filter paper (Soltanpour and Workman, 1979).

7.2. *Potassium Determination:* K in the soil extract is determined by either flame emission, AAS, or by ICP-AES (see 2.3).

8. Standards

8.1. *Potassium Standards* (1,000 mg K L^{-1}): Commercially prepared standard, or weigh 1.9066 g potassium chloride (KCl) in an aliquot of Extraction Reagent (see 6.1) in a 1,000-mL volumetric flask, and bring to volume with Extraction Reagent (see 6.1). Dilute the 1,000 mg K L^{-1} standard with Extraction Reagent (see 6.1) to obtain working standards containing 25, 50, 100, 150, and 250 mg K L^{-1}.

9. Calculations

9.1. mg K kg^{-1} in soil = mg K L^{-1} in extract × 2.

10. Effects of Storage

10.1. Air-drying and storage will not have any significant effect on the level of K. The influence of storage time and temperature is discussed by Houba and Novozamsky (1998).

10.2. The Extraction Reagent (see 6.1) must be stored under mineral oil.

11. Interpretation

11.1. The following tables give an interpretation of the index values for K for the AB-DTPA soil test. These are general guidelines and should be verified under different soil-climate-crop-management combinations.

Index values for Potassium in Soil

Category	K, mg kg^{-1} in soil
Low	0–60
Marginal	60–120
Adequate	>120

12. Comments

12.1. The AB-DTPA test (with a 1:2 soil-solution ratio) index values for K given above regardless of soil texture approximately account for the reduction in K diffusion coefficient with increasing buffering capacity of soils, while NaHCO$_3$ test of Olsen or resin test neglect it due to their much narrower soil:solution ratios. To change AB-DTPA values to Olsen test values multiply the former by 1, 2, and 3, for sandy, sandy loam, and loamy sand soils; loam soils; and clay soils, respectively (Soltanpour, 1991).

7.6 WATER EXTRACTION

1. Principle of the Method

1.1. This method uses water to extract potassium (K), calcium (Ca), magnesium (Mg), and sodium (Na) from soil. A soil-water ratio of 1:5 (v:v) is the one adapted for routine analysis (Bower and Wilcox, 1965; Hesse, 1971; Chapman and Pratt, 1982).

1.2. This method is relatively simple and can serve for quick routine scanning. However, it suffers from the effect of unrealistic figures for Ca and Na due to cation exchange equilibrium shift (Chapman and Pratt, 1982).

2. Range, Sensitivity, and Methods of Analysis

2.1. The range of detection will depend on the particular instrument setup. The range can be extended by diluting the extract.

2.2. The sensitivity will vary with the type of instrument used, wavelength selected, and method of excitation or dissociation.

2.3. The commonly used methods of analysis are flame emission, atomic absorption spectrophotometry (Watson and Isaac, 1990; Wright and Stuczynski, 1996), or inductively coupled plasma emssion spectrometry (ICP-AES) (Soltanpour et al., 1996, 1998).

3. Interferences

3.1. A number of cations and anions which are extracted along with the desired cations will interfere with their determination, particularly Ca by atomic absorption spectrophotometry.

3.2. Known interferences must be eliminated by adding certain cations, such as lanthanum (La), depending on the flame condition used (Watson and Isaac, 1990; Wright and Stuczynski, 1996).

4. Precision and Accuracy

4.1. Repeated analyses of the same soil in the medium concentration range of K, Ca, Mg, and Na will give coefficients of variation of 5% to 10%. A major portion of the variance is related to the heterogeneity of the soil rather than to the extraction or method of analysis.

5. Apparatus

5.1. No. 10 (2-mm opening) sieve.
5.2. 4.25-cm^3 scoop, volumetric.
5.3. Extraction bottle or flask, 50-mL, with stopper.
5.4. Mechanical reciprocating shaker, 180 oscillations per minute.
5.5. Filter funnel, 11 cm.
5.6. Funnel rack.
5.7. Whatman No. 1 filter paper (or equivalent), 11 cm.
5.8. Flame photometer, atomic absorption spectrophotometer (AAS), or inductively coupled plasma emission spectrometer (ICP-AES).
5.9. Analytical balance.
5.10. Volumetric flasks and pipettes as required for preparation of reagents and standard solutions.

6. Reagents (use reagent grade chemicals and pure water)

6.1. *Extraction Reagent:* Pure water.

6.2. *Potassium Standard* (1,000 mg K L^{-1}): Commercially prepared standard, or weigh 1.9080 g potassium chloride (KCl) into a 1,000-mL volumetric flask and bring to volume with water. Prepare working standards by diluting aliquots of the 1,000 mg K L^{-1} standard to cover the anticipated range in concentration to be found in the soil extraction filtrate. Working standards from 5 to 100 mg K L^{-1} should be sufficient for most soils.

6.3. *Calcium Standard* (1,000 mg Ca L^{-1}): Commercially prepared standard, or weigh 2.498 g calcium carbonate (CaCO$_3$) into a 1,000-mL volumetric flask, add 50 mL water, and then add dropwise a minimum volume of concentrated hydrochloric acid (HCl) (approximately 20 mL) to effect complete solution of the CaCO$_3$. Dilute to the mark with water. Prepare working standards by diluting aliquots of the 1,000 mg Ca L^{-1} standard with water to cover the anticipated range in concentration to be found in the soil extraction filtrate. Working standards should be prepared based on the expected concentration in the extract and the range of analytical procedure employed.

6.4. *Magnesium Standard* (1,000 mg Mg L^{-1}): Commercially available standard, or weigh 1.000 g magnesium (Mg) ribbon into a 1,000-mL volumetric flask, dissolve in the minimum volume of dilute HCl, and then bring to volume with water. Prepare the working standards by diluting aliquots of the 1,000 mg Mg L^{-1} standard with pure water to cover the anticipated range in concentration to be found in the soil extraction filtrate. Working standards from 5 to 50 mg Mg L^{-1} should be sufficient for most soils.

6.5. *Sodium Standard* (1,000 mg Na L^{-1}): Commercially prepared standard, or weigh 2.542 g sodium chloride (NaCl) into a 1,000-mL volumetric flask and bring to volume with water. Prepare the working standards by diluting aliquots of the 1,000 mg Na L^{-1} standard with water to cover the anticipated range in concentration to be found in the soil extraction filtrate. Working standards from 1 to 10 mg Na L^{-1} should be sufficient for most soils.

7. Procedure

7.1. *Extraction:* Weigh 5 g or scoop 4.25-cm^3 (see 5.2) of air-dry <10-mesh (2-mm) soil into a 50-mL Extraction Bottle (see 5.3), add 25 mL water, seal the bottle with a stopper, and shake for 30 minutes on a reciprocating shaker (see 5.4). Allow to stand for 15 minutes to let the bulk of the soil settle. Filter the supernatant liquid. Discard the initial filtrate if it is turbid.

7.2. *Analysis:* Ca and Mg in the filtrate can be determined by AAS using either an air-acetylene or a nitrous oxide-acetylene flame (see 2.3), K and Na determined by flame emission spectrophotometry (see 2.3). All four elements can be determined by ICP-AES (see 2.3). Since instruments vary in their operating conditions, no specific details are given here. However, procedures recommended by the manufacturer and described in the operation manual should be followed.

8. Calibration and Standards

8.1. *Working Standards:* Prepare as described in Section 6. If the element concentrations in the extract are found to be outside the range of the instrument or standards, suitable dilutions should be prepared. Dilution should be made to minimize magnification of the error introduced by diluting.

8.2. *Calibration:* Procedures vary with instrumental technique and type. Every precaution should be taken to ensure that the proper procedures are followed, and the manufacturer's recommendations in the operation and calibration of the instrument should be followed.

9. Calculation

9.1. The results are reported as kg ha^{-1} for a 20 cm depth of soil: kg of element ha^{-1} = mg element L^{-1} in extraction filtrate × 10.

9.2. To convert to other units for comparison, see Mehlich (1974).

10. Effects of Storage

10.1. Soils may be stored in an air-dry condition for several months with no effect on element content. The influence of storage time and temperature is discussed by Houba and Novozamsky (1998).

10.2. After extraction, the filtrate should not be stored any longer than 24 hours unless it is treated with acid to prevent bacterial growth and the precipitation of some elements (particularly Ca).

11. Interpretation

11.1. An evaluation of the analytical results, as well as accurate fertilizer recommendations, must be based on field response data conducted under local soil-climate-crop conditions (Peck et al., 1977; Brown, 1987; Dahnke and Olson, 1990; Haby et al., 1990; Black, 1993; Mikelsen and Camberato, 1995; Maynard and Hochmuth, 1997; Lukwick, 1998; Reid, 1998).

7.7 REFERENCES

Aitken, R. L. and B. J. Scott. 1999. Magnesium, In: *Soil Analysis: An Interpretation Manual,* K. I. Peverill, L. A. Sparrow, and D. J. Reuter (eds.). Collingwood, Australia: CSIRO Publishing, 255–262.

Anon. 1994a. Extractable phosphorus, potassium, sulfate-sulfur, zinc, iron, manganese, copper, nitrate: Ammonium bicarbonate-DTPA method, In: *Plant, Soils and Water Reference Methods for the Western Region,* Western Regional Extension Publication WREP 125, R. G. Gavlak, D. A. Horneck, and R. O. Miller (eds.). Fairbanks, AK: University of Alaska, 27–28.

Anon. 1994b. Extractable phosphorus, potassium, calcium, magnesium, sodium, iron, manganese, and copper: Mehlich-3 method, In: *Plant, Soils, and Water Reference Methods for the Western Region,* Western Regional Extension Publication WREP 125, R. G. Gavlak, D. A. Horneck, and R. O. Miller (eds.). Fairbanks, AK: University of Alaska, 25–26.

Anon. 1994c. Extractable calcium, magnesium, potassium, and sodium: Ammonium acetate method, In: *Plant, Soils, and Water Reference Methods for the Western Region,* Western Regional Extension Publication WREP 125, R. G. Gavlak, D. A. Horneck, and R. O. Miller (eds.). Fairbanks, AK: University of Alaska, 31–32.

Anon. 1996. Potassium, *Soil Fertility Manual,* Chapter 5. Norcross, GA: Potash & Phosphate Institute, 5:1–9.

Anon. 1998. *Soil Test Levels in North America: Summary Update.* PPI/PPIC/FAR Technical Bulletin 1998–3. Norcross, GA: Potash & Phosphate Institute.

Beegle, D. B. and T. C. Oravec. 1990. Comparison of field calibrations for Mehlich-III P and K with Bray P1 and ammonium acetate K for corn. *Commun. Soil Sci. Plant Anal.* 21:1025–1036.

Bates, T. E. and J. E. Richards. 1993. Available potassium, In: *Soil Sampling and Methods of Analysis,* M. R. Carter (ed.). Boca Raton, FL: CRC Press, 59–64.

Black, C. A. 1993. *Soil Fertility Evaluation and Control.* Boca Raton, FL: Lewis Publishers.

Bower, C. A. and L. V. Wilcox, 1965. Soluble Salts, In: *Methods of Soil Analysis,* Part 2. Agronomy No. 9, C. A. Black (ed.). Madison, WI: American Society of Agronomy, 935–945.

Brown, J. R. (ed.). 1987. *Soil Testing: Sampling, Correlation, Calibration, and Interpretation,* SSSA Special Publication 21. Madison, WI: Soil Science Society of America.

Bruce, R. C. 1999. Calcium, In: *Soil Analysis: An Interpretation Manual,* K. I. Peverill, L. A. Sparrow, D. J. Reuter (eds.). Collingwood, Australia: CSIRO Publishing, 247–254.

Chapman, H. D. and P. Pratt. 1982. *Method of Analysis for Soils, Plants, and Water.* Priced Publication 4034, Berkeley: University of California, Division of Agricultural Sciences.

Dahnke, W. C. and R. A. Olson. 1990. Soil test correlation, calibration, and recommendation, In: *Soil Testing and Plant Analysis,* 3rd ed., SSSA Book Series No. 3. R. L. Westerman (ed.). Madison, WI: Soil Science Society of America, 45–71.

Flannery, R. L. and D. K. Markus. 1972. Determination of phosphorus, potassium, calcium and magnesium simultaneously in North Carolina, ammonium acetate, and Bray P1 soil extracts by AutoAnalyzer, In: *Instrumental Methods for Analysis of Soils and Plant Tissue,* L. M. Walsh (ed.). Madison, WI: Soil Science Society of America, 97–112.

Gourley, C. J. P. 1999. Potassium, In: *Soil Analysis: An Interpretation Manual,* K. I. Peverill, L. A. Sparrow, D. J. Reuter (eds.). Collingwood, Australia: CSIRO Publishing, 229–245.

Greweling, T. and M. Peech. 1965. *Chemical Soil Tests.* Cornell Agricultural Experiment Station Bulletin 960.

Haby, V. A., M. P. Russelle, and E. O. Skogley. 1990. Testing soils for potassium, calcium, and magnesium, In: *Soil Testing and Plant Analysis,* 3rd ed., SSSA Book Series No. 3. R. L. Westerman (ed.). Madison, WI: Soil Science Society of America, 181–227.

Hanlon, E. A. and G. V. Johnson. 1984. Bray/Kurtz, Mehlich-III, AB/DTPA, and ammonium acetate extractions for P, K, and Mg in four Oklahoma soils. *Commun. Soil Sci. Plant Anal.* 15:277–294.

Hatfield, A. L. 1972. Soil test reporting: A nutrient index system. *Commun. Soil Sci. Plant Anal.* 3:425–436.

Helmke, P. A. and D. L. Sparks. 1996. Lithium, sodium, potassium, rubidium, and calcium, In: *Methods of Soil Analysis, Part 3, Chemical Methods,* SSSA Book Series No. 5, D. L. Sparks (ed.). Madison, WI: Soil Science Society of America, 551–574.

Hesse, P. R. 1971. *A Textbook of Soil Chemical Analysis.* New York: Chemical Publishing Co.

Houba, V. J. G. and I. Novozamsky. 1998. Influence of storage time and temperature of air-dried soils on pH and extractable nutrients using 0.01 M $CaCl_2$. *Fresenicus J. Anal. Chem.* 360:362–365.

Isaac, R. A. and J. B. Jones, Jr. 1970. AutoAnalyzer systems for the analysis of soil and plant tissue extracts. *Adv. Auto. Anal.* 2:57–64.

Isaac, R. A. and J. B. Jones, Jr. 1971. AutoAnalyzer systems for the analysis of soil and plant tissue extracts, Terrytown, NY: *Technicon International Congress,* 57–64.

Jackson, M. L. 1958. *Soil Chemical Analysis.* Englewood Cliffs, NJ: Prentice-Hall.

Jones, Jr., J. B. 1973. Soil testing in the United States. *Commun. Soil Sci. Plant Anal.* 4:307–322.

Jones, Jr., J. B. 1998a. *Plant Nutrition Manual.* Boca Raton, FL: CRC Press.

Jones, Jr., J. B. 1998b. Soil test methods: Past, present, and future. *Commun. Soil Sci. Plant Anal.* 29:1543–1552.

Ludwick, A. E. 1998. *Western Fertilizer Handbook.* 2nd ed. Danville, IL: Interstate Publishers, Inc.

Lunt, H. A., C. L. W. Swanson, and H. G. M. Jacobson. 1950. *The Morgan Soil Testing System.* New Haven, CT. Connecticut Agricultural Experiment Station. Bulletin 541.

Maynard, D. N., and G. J. Hochmuth. 1997. *Knott's Handbook for Vegetable Growers,* 4th ed. New York: John Wiley & Sons.

Mehlich, A. 1953. *Determination of P, Ca, Mg, K, Na and NH_4* (mimeo). Raleigh, NC: North Carolina Soil Test Division.

Mehlich, A. 1972. Uniformity of expressing soil test results: A case for calculating results on a volume basis. *Commun. Soil Sci. Plant Anal.* 3:417–424.

Mehlich, A. 1974. Uniformity of soil test results as influenced by extractants and soil properties, In: *Proceedings Seventh International Colloquium Plant Analysis and Fertilizer Problems,* J. Wehrmann (ed.). Hanover, Germany; 295–305.

Mehlich, A. 1978. New extractant for soil test evaluation of phosphorus, potassium, magnesium, calcium, sodium, manganese, and zinc. *Commun. Soil Sci. Plant Anal.* 9:477–492.

Mehlich, A. 1984. Mehlich 3 soil test extractant: A modification of Mehlich 2 extractant. *Commun. Soil Sci. Plant Anal.* 15:1409–1416.

McLean, E. O. 1977. Contrasting concepts of soil test interpretation: Sufficiency levels of available nutrients versus basic cation saturation ratios, In: *Soil Testing: Correlating and Interpreting the Analytical Results,* ASA Special Publication No. 29, T. R. Peck et al. (eds.). Madison, WI: American Society of Agronomy, 39–54.

Mikkelsen, R. L. and J. J. Camberato. 1995. Potassium, sulfur, lime, and micronutrient fertilizers, In: *Soil Amendments and Environmental Quality,* J. E. Rechcigl (ed.). Boca Raton, FL: Lewis Publishers, 109–137.

Morgan, M. F. 1941. *Chemical Diagnosis by the Universal Soil Testing System,* Bulletin 450. Connecticut New Haven, CT: Agricultural Experiment Station.

Nelson, W. L., A. Mehlich, and E. Winters. 1953. The development, evaluation and use of soil tests for phosphorus availability, In: *Soil and Fertilizer Phosphorus.* Agronomy No. 4, W. H. Pierre and A. G. Norman (eds.). Madison, WI: American Society of Agronomy, 153–188.

Peck, T. R. and J. T. Cope, Jr., and D. A. Whitney (eds.). 1977. *Soil Testing: Correlating and Interpreting the Analytical Results,* ASA Special Publication 29. Madison, WI: American Society of Agronomy.

Peverill, K. I., L. A. Sparrow, and D. J. Reuter (eds.). 1999. *Soil Analysis: An Interpretation Manual.* Collingwood, Australia: CSIRO Publishing.

Prianischnikov, D. N. 1913. *Landw. vers. Sta.* 79–80:667.

Rechcigl, J. E., G. G. Payne, and C. A. Sanchez. 1992. Comparison of various soil drying techniques on extractable nutrients. *Commun. Soil Sci. Plant Anal.* 23: 2347–2363.

Reid, K. 1998. *Soil Fertility Handbook.* Toronto: Ministry of Agriculture, Food and Rural Affairs, Queen's Printer for Ontario.

Schmisek, M. E., L. J. Cihacek, and L.J. Swenson. 1998. Relationships between the Mehlich-III soil test extraction procedure and standard soil test methods in North Dakota. *Commun. Soil Sci. Plant Anal.* 29:1719–1729.

Schollenberger, C. J. and R. H. Simon. 1945. Determination of exchange capacity and exchangeable bases in soil-ammonium acetate method. *Soil Sci.* 59:13–24.

Sen Tram, T. and R. R. Simard. 1993. Mehlich III-extractable elements, In: *Soil Sampling and Methods of Analysis,* M. R. Carter (ed.). Boca Raton, FL: CRC Press, 43–49.

Simard, R. R. 1993. Ammonium acetate-extractable elements, In: *Soil Sampling and Methods of Analysis,* M. R. Carter (ed.). Boca Raton, FL: CRC Press, 39–42.

Sims, J. T. 1989. Comparison of Mehlich 1 and Mehlich 3 extractants for P, K, Ca, Mg, Mn, Cu, and Zn in Atlantic coastal plain soils. *Commun. Soil Sci. Plant Anal.* 20:1707–1726.

Soltanpour, P. N. 1991. Determination of nutrient element availability and elemental toxicity by the AB-DTPA soil test and ICPS. *Adv. Soil Sci.* 16:165–190.

Soltanpour, P. N. and A P. Schwab. 1977. A new soil test for simultaneous extraction of macro- and micronutrients in alkaline soils. *Commun. Soil Sci. Plant Anal.* 8:195–207.

Soltanpour, P. N. and S. M. Workman. 1979. Modification of NH_4HCO_3-DTPA soil test to omit carbon black. *Commun. Soil Sci. Plant Anal.* 10:1411–1420.

Soltanpour, P. N., A. Khan, and A. P. Schwab. 1979. Effect of grinding variables on the NH_4HCO_3-DTPA soil test values for Fe, Zn, Mn, Cu, P, and K. *Commun. Soil Sci. Plant Anal.* 10:903–909.

Soltanpour, P. N., G. W. Johnson, S. M. Workman, J. B. Jones, Jr., and R. O. Miller. 1996. Inductively coupled plasma emission spectrometry and inductively coupled plasma-mass spectrometry, In: *Methods of Soil Analysis, Part 3, Chemical Methods,* SSSA Book Series No. 5. D.L. Sparks (ed.). Madison, WI: Soil Science Society of America, 91–139.

Soltanpour, P. N., C. W. Johnson, S. M. Workman, J. B. Jones, Jr., and R. O. Miller. 1998. Advances in ICP emission and ICP mass spectrometry. *Adv. Agron.* 64:28–113.

Sparks, D. L. 1987. Potassium soil dynamics in soils. *Adv. Soil Sci.* 6:1–63.

Suarez, D. L. 1996. Beryllium, magnesium, calcium, strontium, and barium, In: *Methods of Soil Analysis, Part 3, Chemical Methods.* SSSA Book Series No. 5, D. L. Sparks (ed.). Madison, WI: Soil Science Society of America, 575–601.

Tucker, M. R. 1992. Determination of potassium, calcium, magnesium, and sodium by Mehlich 3 extraction, In: *Reference Soil and Media Diagnostic Procedures for the Southern Region of the United States.* Southern Cooperative Series Bulletin No. 374, S. J. Donohue (ed.). Blacksburg, VA: Virginia Agricultural Experiment Station, 9–12.

Warncke, D., and J. R. Brown. 1998. Potassium and other basic cations, In: *Recommended Chemical Soil Test Procedures for the North Central Region.* North Central Regional Publication No. 221 (revised), J. R. Brown (ed.). Columbia, MO: Missouri Agricultural Experiment Station SB 1001, University of Missouri, 31–33.

Watson, M. E., and R. A. Isaac. 1990. Analytical instruments for soil and plant analysis. In: *Soil Testing and Plant Analysis,* 3rd ed., SSSA Book Series No. 3. R. L. Westerman (ed.). Madison, WI: Soil Science Society of America, 691–740.

Wolf, A. M. and D. Beegle. 1995. Recommended soil tests for macronutrients: Phosphorus, potassium, calcium and magnesium, In: *Recommended Soil Testing Procedures for the Northeastern United States.* Northeast Regional Publication No. 493 (revised), J. T. Sims and A. M. Wolf (eds.). Newark, DE: Agricultural Experiment Station, University of Delaware, 30–38.

Wolf, B. 1982. An improved universal extracting solution and its use for diagnosing soil fertility. *Commun. Soil Sci. Plant Anal.* 13:1005–1033.

Wright, R. J. and T. I. Stuczynski. 1996. Atomic absorption and flame emission spectrometry, In: *Methods of Soil Analysis, Part 3, Chemical Methods,* SSSA Book Series No. 5. D. L. Sparks (ed.). Madison, WI: Soil Science Society of America, 65–90.

8 Micronutrients (Boron, Copper, Iron, Manganese, and Zinc)

The micronutrients, boron (B), chlorine (Cl), copper (Cu), iron (Fe), manganese (Mn), molybdenum (Mo), and zinc (Zn), are the seven elements essential for plants at requirement levels of less than 0.10% (Epstein, 1972; Glass, 1989). The micronutrient Cl is covered in Chapter 12, and a test procedure for Mo is not given in this handbook. Since the micronutrients Cu, Mn, and Zn are also classed as heavy metals, some of these elements are included in Chapter 9.

In general, most of the micronutrient soil tests are limited in their application and are frequently associated with certain crop and soil characteristics (Martens and Lindsay, 1990, Johnson and Fixen, 1990; Pais and Jones, 1997; Jones, 1998a), considerations that are essential for the interpretation of a micronutrient soil test. A general discussion of the micronutrients is given in the Potash & Phosphate Institute's *Soil Fertility Manual* (Anon., 1996).

Shuman (1991) has described the chemical forms of the micronutrients that exist in soil, Harter (1991) their adsorption-desorption reactions, Lindsay (1991) inorganic equilibria, Stevenson (1991) organic matter reactions, and Moraghan and Mascagni (1991) the environmental and soil factors affecting deficiencies and toxicities, factors that influence micronutrient uptake and utilization by crops. The geochemistry of the trace elements is described in the book edited by Adriano (1992). Soil analysis (test) methods for all the micronutrients are given by Sims and Johnson (1991) and Sims (1995), and for B by Wear (1965), Johnson and Fixen (1990), Gupta (1990), Isaac (1992), Anon. (1994d), Keren (1996), Watson (1998), and Bell (1999); Cu by Martens and Lindsay (1990), Anon. (1994abc) and Brennan and Best (1999); Cu and Zn by Johnson (1992), Tucker (1992), Reed and Martens (1996), and Anon. (1994abc); Fe by Martens and Lindsay (1990), Ross and Wang (1990), Anon. (1994abc) Loeppert and Inskeep (1996), and MacFarlane (1999); Mn by Ross and Wang (1990), Martens and Lindsay (1990), Tucker (1992), Johnson (1992), Anon. (1994abc), Gambrell (1996), and Uren (1999); Zn by Viets and Boawn (1965), Nelson et al. (1959), Wear and Evans (1968), Martens and Lindsay (1990) and Armour and Brennan (1999); Cu, Fe, Mn, and Zn by Viets and Lindsay (1973), Marten and Lindsay (1990), and Whitney (1998); B, Cu, Mn, and Zn by Sims (1995); and Cu, Mn, and Zn by Tucker (1992). The commonly used micronutrient soil testing procedures were recently published by Jones (1998b).

General tests on the micronutrients have been written by Adriano (1986), Kabata-Pendias and Pendias (1995), and Pais and Jones (1998). Correlation and calibration techniques for these elements can be found in the articles by Peck et al.

(1977) and Sims and Johnson (1991), while specific techniques when dealing with a micronutrient are given in the presentations by Alley et al. (1972), Cox and Wear (1977), and Armour and Brennan (1999) for Zn, Makarim and Cox (1983) and Brennan and Best (1999) for Cu, and Mascagni and Cox (1985) and Uren (1999) for Mn, McFarlane (1999) for Fe, and Bell (1999) for B. General recommendations for the use and application rates for the micronutrients can be found in the books by Halliday and Trenkel (1992) and Marynard and Hochmuth (1997), and in the articles by Martens and Westerman (1991), Kabata-Pendias and Adriano (1995), and Mikkelsen and Camberato (1995).

8.1 HOT WATER BORON EXTRACTION

1. Principle of the Method

1.1. This method determines the amount of available soil boron (B). It was first proposed by Berger and Truog (1940) and described in detail by Wear (1965), Bingham (1982), Johnson and Fixen (1990), Gupta (1990), Anon. (1994d), Keren (1996), Isaac (1992), Watson (1998), and Bell (1999). Various modifications of the technique have been studied by Gupta (1967) and Odom (1980). Wolf (1971, 1974) has developed a different extraction method using azomethine-H as the color development reagent. However, hot water extraction is a method in common use, although other methods have been proposed (Johnson and Fixen, 1990). Shuman et al. (1992) found that B extracted by either the Mehlich No. 1 and No. 3 extractants compared favorably with that extracted by hot water.

2. Range, Sensitivity, and Methods of Analysis

2.1. Boron can be extracted and determined in soil concentrations from less than 1 to 10 mg L^{-1} without dilution. The upper limit may be extended by diluting the extract prior to spectrophotometric determination.

2.2. The technique for extraction involves boiling the soil in water for a specified time period. The effect of time and boiling technique has been studied by Gupta (1967) and Odom (1980). A mechanized technique has been developed by John (1973).

2.3. The commonly used method of analysis is UV-VIS spectrophotometry (Watson and Isaac, 1990; Wright and Stuczynski, 1996), although B can also be determined by inductively coupled plasma emission spectrometry (ICP-AES) (Soltanpour et al., 1996, 1998).

3. Interferences

3.1. Obtaining a clear filtrate after boiling may be a problem. The use of double filter paper (see 7.1) is helpful. Some have proposed boiling soil in a 0.1% solution of calcium chloride ($CaCl_2 \cdot 2H_2O$) to help clear the filtrate.

3.2. Interferences may occur in the color development when various color reagents (Berger and Truog, 1944; Gupta, 1967; Wear, 1965; Reisenauer et al., 1973) other than those proposed here are used. With the use of azomethine-H, interferences are minimal (Wolf, 1971, 1974).

Micronutrients (Boron, Copper, Iron, Manganese, and Zinc)

4. Precision and Accuracy

4.1. Variance is essentially a factor related to the heterogeneity of the soil rather than to the extraction or colorimetric procedures. However, for the azomethine-H procedure, careful mixing of this reagent with the soil extractant is important (Wolf, 1974).

5. Apparatus

5.1. No. 10 (2-mm) sieve.
5.2. 8.5-cm^3 scoop, volumetric.
5.3. Refluxing apparatus (boron-free glassware), 250-mL capacity.
5.4. Hot plate.
5.5. Filter funnel, 11 cm and funnel rack.
5.6. Whatman No. 42 filter paper or equivalent, 11 cm.
5.7. UV-VIS spectrophotometer set at 430 nm.
5.8. Spectrophotometric tube or cuvet.
5.9. Volumetric flasks and pipettes as required for preparation of reagents, standard solutions, and color development.
5.10. Analytical balance.

6. Reagents (use reagent grade chemicals and pure water)

6.1. *Azomethine-H Reagent:* Weigh 0.9 g azomethine-H and 2 g ascorbic acid into 10 mL water and follow with gentle heating in a water bath. When dissolved, dilute to 100 mL with water. If the solution is turbid, reheat in the water bath until it is clear. The reagent will last 14 days when refrigerated.

6.2. *Buffer Masking Reagent:* Weigh 250 g ammonium acetate ($NH_4C_2H_3O_2$), 25 g tetrasodium salt of (ethylenedinitrilo) tetraacetic acid, and 10 g disodium salt of nitrilotriacetic acid in 400 mL water. Slowly add 125 mL acetic acid ($HC_2H_3O_2$).

6.3. *Boron Standard* (100 mg B L^{-1}): Commercially prepared standard, or weigh 0.5716 g boric acid (H_3BO_3) into a 1,000-mL volumetric flask and bring to volume with water.

7. Procedure

7.1. *Extraction:* Weigh 10 g or scoop 8.5-cm^3 (see 5.2) of air-dry < 10-mesh (2-mm) soil into a refluxing flask and add 20 mL water. Assemble the refluxing apparatus and place the flasks on the hot plate. Bring to a boil and boil 10 minutes. Filter through double filter paper (see 5.6) and collect the filtrate for B determination.

7.2. *Color Development:* Pipette 4 mL extractant into a spectrophotometer cuvet, add 1 mL Buffer Masking Reagent (see 6.2), and 1 mL Azomethine-H Reagent (see 6.1). After adding the Azomethine-H Reagent, mix the solution immediately. Cap and invert the cuvet and let stand for one hour. Using an UV-VIS spectrophotometer, read the transmittance (% T) at 430 nm with the blank being water.

8. Calibration and Standards

8.1. *Working Boron Standards:* With the Boron Standard (see 6.3), prepare five Working Standards giving 0.125, 0.25, 0.5, 1.0, and 2.0 mg B L^{-1}. Use 4 mL aliquots of each standard and carry through the color development (see 7.2).

8.2. Calibration Curve: On semi-log graph paper, plot the percent transmittance on the logarithmic scale versus mg B L^{-1} on the linear scale.

9. Calculation

9.1. The results are reported as mg B kg^{-1} in soil mg B kg^{-1} in soil = mg B L^{-1} in the extractant × 2. If the extraction filtrate is diluted, the dilution factor must be applied.

10. Effects of Storage

10.1. Soils may be stored in an air-dry condition for several months with no effect on extractable B. The influence of time and temperature is discussed by Houba and Novozamsky (1998).

10.2. After extraction, the extraction solution containing B should not be stored any longer than 24 hours.

11. Interpretation

11.1. Accurate fertilizer recommendations for B must be based on known field responses based on local soil-climate-crop conditions (Berger and Truog, 1940, 1944; Gupta, 1967; Wear, 1965; Reisenauer et al., 1973; Johnson and Fixen, 1990; Dahnke and Olson, 1990; Jones, 1998). For most soils and crops, the amount of B extracted should be interpreted as follows:

Category	mg B kg^{-1} in soil
Insufficient	<1.0
Adequate	1.0–2.0
High	2.1–5.0
Excessive	>5.0

8.2 MEHLICH NO. 1 ZINC EXTRACTION

1. Principle of the Method

1.1. This method for determining extractable zinc (Zn) has been evaluated only on soils which have cation exchange capacities of less than 10 milliequivalents per 100 grams, are acid (pH less than 7.0) in reaction, and are relatively low (less than 5%) in organic matter content. Its suitability for use on alkaline or organic soils has not been determined. This method is described in some detail by Perkins (1970) for use with the sandy coastal plain soils of the southeastern United States.

1.2. The use of Mehlich No. 1 as an extraction reagent for cations and phosphorus (P) was first reported by Mehlich (1953), and later classified as a P extraction reagent by Nelson et al. (1953), as the North Carolina Double Acid Method (renamed Mehlich No. 1). Wear and Evans (1968) compared Mehlich No. 1 versus 0.1 N hydrochloric acid (HCl) and EDTA as a Zn extraction reagent on 12 soils and found Mehlich No. 1-extractable Zn to correlate more closely with Zn uptake by corn and sorghum plants. Alley et al. (1972), developed a prediction equation for

Micronutrients (Boron, Copper, Iron, Manganese, and Zinc) 121

Zn response under field conditions using Mehlich No. 1-extractable Zn. The equation was improved considerably by taking into consideration soil pH and Mehlich No. 1-extractable P.

2. Range, Sensitivity, and Methods of Analysis

2.1. Zinc can be extracted and determined in soil concentrations from 0.4 to 8.0 mg Zn L^{-1} without dilution. The range and upper limits may be extended by diluting the extraction filtrate prior to analyses.

2.2. Sensitivity will vary with the type of instrument used, wavelength, and method of excitation.

2.3. The commonly used method of analysis is atomic absorption spectrophotometry (AAS) (Watson and Isaac, 1990; Wright and Stuczynski, 1996) or by inductively coupled plasma emission spectrometry (ICP-AES) (Soltanpour et al., 1996, 1998).

3. Interferences

3.1. Known interferences and compensation for changing characteristics of the extract to be analyzed must be acknowledged. However, no serious interferences have been reported.

3.2. The method of soil drying has a significant effect on the amount of Zn extracted (Rechcigl et al., 1992).

4. Precision and Accuracy

4.1. The major source of variance in the extraction is the heterogeneity of the soil sample itself. Repeated analyses of the same soil sample will give a coefficient of variation of approximately 10%.

5. Apparatus

5.1. No. 10 (2-mm opening) sieve.
5.2. 4-cm^3 scoop, volumetric.
5.3. 50-mL extraction bottle or flask, with stopper.
5.4. Mechanical reciprocating shaker, 180 oscillations per minute.
5.5. Filter funnel, 11 cm.
5.6. Funnel rack.
5.7. Acid-washed filter paper, Whatman No. 42, 11 cm or equivalent.
5.8. Atomic absorption spectrophotometer (AAS) or inductively coupled plasma emission spectrometer (ICP-AES).
5.9. Analytical balance.
5.10. Volumetric flasks and pipettes as required for preparation of reagents and standard solutions.

6. Reagents (use reagent grade chemicals and pure water)

6.1. *Extraction Reagent* (0.05 N HCl in 0.025 N H$_2$SO$_4$): Pipette 4.3 mL conc. hydrochloric acid (HCl) and 0.7 mL conc. sulfuric acid (H$_2$SO$_4$) into a 1,000-mL volumetric flask and bring to volume with water.

6.2. *Zinc Standard* (1,000 mg Zn L^{-1}): Commercially prepared standard, or weigh 4.3478 g zinc sulfate (ZnSO$_4$·7H$_2$O) into a 1,000-mL volumetric flask and bring to volume with Extraction Reagent (see 6.1). Prepare working standards by diluting aliquots of the 1,000 mg Zn L^{-1} standard with Extraction Reagent (see 6.1) to cover the anticipated range in concentration to be found in the soil extraction filtrate. Working standards from 0 to 3 mg Zn L^{-1} in solution should be sufficient for most soils.

7. Extraction Procedure and Analysis

7.1. *Extraction:* Weigh 5 g or scoop 4-cm^3 air-dry < 10-mesh (2-mm) soil into an acid-washed 50-mL Extraction Bottle (see 5.3), add 20 mL Extraction Reagent (see 6.1), and shake for five minutes on a reciprocating shaker (see 5.4). Filter and collect the filtrate.

7.2. *Blank:* An Extraction Bottle (see 5.3) without soil but added Extracting Reagent (see 6.1) should be carried through the shaking and filtering procedure to serve as the *blank,* substracting the Zn determined from the soil analysis result.

7.3. *Analysis:* Zn in the filtrate can be determined by either AAS or ICP-AES (see 2.3). Since instruments vary in their operating conditions, no specific details are given.

8. Calibration and Standards

8.1. *Working Standards:* Working standards should be prepared as described in Section 6.2. If concentrations are found outside the range of the instrument or standards, suitable dilutions should be prepared starting with a 1:1 soil-to-Extraction Reagent dilution. Dilution should be made so as to minimize the magnification of errors introduced by diluting.

8.2. *Calibration:* Procedures vary with instrument techniques and the type of instrument. Every precaution should be taken to ensure that the proper procedures are followed and that the manufacturer's recommendations in the operation and calibration of the instrument are used.

9. Calculation

9.1. For results reported as lb Zn A^{-1} in the soil, multiply the mg Zn L^{-1} in the filtrate x 8 (substract the *blank* Zn value from the soil assay result before multiplying the dilution factor). To convert to other units, see Mehlich (1972, 1973).

10. Effects of Storage

10.1. Soils may be stored in an air-dried condition for several months with no effects on the extractable Zn content. The influence of time and temperature is discussed by Houba and Novozamsky (1998).

10.2. After extraction, the filtrate should not be stored longer than 24 hours without refrigeration or other treatment to prevent bacterial growth.

11. Interpretation

11.1. An evaluation of the results as well as Zn recommendations must be based on field response data conducted under local soil-climate-crop conditions (Viets et al.,

1973; Dahnke and Olson, 1990; Jones 1998). Interpretative data that would be applicable to the southeastern United States are given by Alley et al. (1972), Perkins (1970), Cox and Wear (1977), and Martens and Lindsay (1990).

11.2. The critical soil test Zn level for corn as interpreted by Cox and Wear (1977) is 0.8 mg Zn kg^{-1}. The probability of a corn yield response to Zn fertilization on soils testing below this value would be high. This critical level may not apply to extremely high CEC soils, high P containing soils or very acid soils.

8.3 MEHLICH NO. 3 EXTRACTION

1. Principle of the Method

1.1. The extraction and determination of boron (B), copper (Cu), manganese (Mn), and zinc (Zn) by this procedure is applicable across a wide range of soil properties ranging in reaction from acid to basic (Anon., 1994b). Although the method was correlated with established extractants from several regions and critical levels were established, the specific critical levels should be based on local soil, crop, and climatic conditions. Good correlations were obtained between Mehlich No. 1 and Mehlich No. 3 for the micronutrients, B (Shuman et al., 1992), Cu, Mn, and Zn (Sims, 1989), and Mehlich No. 2 (Mehlich, 1978) and Mehlich No. 3 for Mn and Zn (Mehlich, 1984), even though the mean values were not the same (Schmisek et al., 1989; Tucker, 1992). Critical Zn levels for this extractant have been given by Mehlich (1984).

2. Range, Sensitivity, and Methods of Analysis

2.1. Cu, Mn, and Zn can be extracted and determined without dilution in soil concentrations of 4.0, 20, and 4.0 mg dm^{-3}, respectively. This equates to 8.0, 40, and 8.0 kg ha^{-1} for Cu, Mn, and Zn, respectively. Higher concentrations can be determined with appropriate dilutions or by using instrumentation equipped with a 3-standard microprocessor. The method was developed using atomic absorption spectrophotometry (AAS) at a 1:10 soil-to-solution ratio for Cu, Mn, and Zn (Mehlich, 1984), and UV-VIS spectrophotometry for B.

2.2. Sensitivity will vary with the type of instrument, wavelength selection, and method of excitation. Watson and Isaac (1990) and Wright and Stuczynski (1996) have, described these parameters for the ASS and UV-VIS spectrophotometry procedures, and Soltanpour et al. (1996, 1998) for the ICP-AES procedure.

3. Precision and Accuracy

3.1. Repeated analyses of an internal check sample from 20 extractions by the NCDA Soil Testing Laboratory gave variances of 9.69%, 10.82%, and 9.44% for Mn, Zn, and Cu, respectively. The mean values were 5.42, 2.29, and 1.71 mg dm^{-3} of Mn, Zn, and Cu, respectively. The variance is essentially a factor related to sample heterogeneity rather than to measurement, extraction, or method of analysis.

4. Interferences and Contamination

4.1. There are no known interferences.

4.2. The possibility of contamination between samples or from external sources (extraction vials, filter funnels and washing apparatus) should be recognized. Precautions should be taken to avoid the use of extraction vials which contain micronutrient impurities. Certain plastic bottles are also charged and can retain Cu and Zn from previous extractions. Consequently, all laboratory apparatus must be washed with a reagent capable of displacing absorbed micronutrients. The rinsing solution used in this procedure is described below (see 7.1).

5. Apparatus

5.1. No. 10 (2-mm) stainless steel sieve.
5.2. 5-cm^3 scoop (volumetric) and Teflon-coated leveling rod.
5.3. 100-mL extraction bottles, preferably plastic.
5.4. Reciprocating shaker (200, 4-cm reciprocations per minute).
5.5. Whatman No. 1 (or equivalent) filter paper, 11 cm.
5.6. Plastic filter funnels, 11 cm.
5.7. Funnel rack.
5.8. Vials, polystyrene plastic, 25-mL capacity, for sample collection.
5.9. Automatic dispenser for extractant, 25-mL capacity.
5.10. UV-VIS spectrophotometer, atomic absorption spectrophotometer (AAS) or inductively coupled plasma emission spectrometer (ICP-AES).
5.11. Volumetric flasks and pipettes as required for preparation of reagents and standard solutions.
5.12. Analytical balance.

6. Reagents (use reagent grade chemicals and pure water)

6.1. *Extraction Reagent* [0.2 N acetic acid (CH$_3$COOH); 0.25 N ammonium nitrate (NH$_4$NO$_3$); 0.015 N ammonium fluoride (NH$_4$F); 0.013 N nitric acid (HNO$_3$); 0.001 M EDTA]: see 6.2 and 6.3 for the mixing procedure for preparing the extraction reagent.

6.2. *Ammonium Fluoride—EDTA Stock Reagent:* Add approximately 600 mL water to a 1,000-mL volumetric flask, add 138.9 g NH$_4$F, and dissolve. Then add 73.05 g EDTA. Dissolve the mixture and bring to volume with water.

6.3. *Final Extraction Reagent Mixture:* Add approximately 3,000 mL water to a 4,000-mL volumetric flask, add 80 g NH$_4$NO$_3$, and dissolve. Add 16 mL NH$_4$F-EDTA Stock Reagent (see 6.2) and mix well. Add 46 mL glacial CH$_3$COOH and 3.28 mL conc. nitric acid (HNO$_3$), then bring to volume with water and mix thoroughly. The final pH should be 2.5 ± 0.1.

6.4. *Boron Standard* (100 mg B L^{-1}): Commercially prepared standard, or weigh 0.5716 g boric acid (H$_3$BO$_3$) into a 1,000-mL volumetric flask and bring to volume with Extraction Reagent (see 6.3).

6.5. *Boron Working Standards:* Dilute 2.5, 5.0, 7.5, and 10 mL Boron Standard (see 6.4) to 1,000 mL with Extraction Reagent (see 6.3), corresponding to 0.25. 0.50, 0.75, and 1.00 mg B L^{-1}, respectively.

6.6. *Procedures and Calculations:* With the 1.0 mg B L^{-1} Boron Working Standard (see 6.5), set the instrument reading to 100. Soil extracts can be read directly with appropriate dilutions when instrument readings exceed 100. Instrument readout × 0.2 = mg B dm^{-3} of soil; mg B dm^{-3} × 2 = kg B ha^{-1} and kg B ha^{-1} × 0.891 = lbs B A^{-1}. These calculations are based on a volume of soil to a depth of 20 cm. For the rationale, see Mehlich (1972, 1973).

6.7. *Manganese Standard* (20 mg Mn L^{-1}): Dilute 20 mL commercially prepared Manganese Reference Standard (1,000 mg Mn L^{-1}) to 1,000 mL with Extraction Reagent (see 6.3).

6.8. *Manganese Working Standards:* Dilute 25, 50, 75, and 100 mL Manganese Standard (see 6.6) to 1,000 mL with Extraction Reagent (see 6.3), corresponding to 0.5, 1.0, 1.5, and 2.0 mg Mn L^{-1}, respectively. Following the manufacturer's guidelines, set the instrument at zero with Extraction Reagent (see 6.3). Using the 2.0 mg Mn L^{-1} standard, set the instrument reading to 100. Intermediate standards can be used to check linearity. Higher concentration ranges can be used with an atomic absorption spectrophotometer equipped with a 3-standard microprocessor.

6.9. *Procedure and Calculations:* With the 2.0 mg Mn L^{-1} Manganese Working Standard (see 6.7), set the instrument reading to 100. Soil extracts can be read directly with appropriate dilutions when instrument readings exceed 100. Instrument readout × 0.2 = mg Mn dm^{-3} of soil; mg Mn dm^{-3} × 2 = kg Mn ha^{-1} and kg Mn ha^{-1} × 0.891 = lbs Mn A^{-1}. These calculations are based on a volume of soil to a depth of 20 cm. For the rationale, see Mehlich (1972, 1973).

6.10. *Zinc and Copper Standards* (4 mg Zn, Cu L^{-1}): Dilute 100 mL commercially prepared reference standard (1,000 mg Zn and Cu L^{-1} to 1,000 mL with Extraction Reagent (see 6.3). From this mixture, dilute 40 mL to 1,000 mL with Extraction Reagent (see 6.3) for corresponding concentrations of 4 mg Zn and Cu L^{-1}. These standards can be prepared separately or in combination with each other, depending on preference.

6.11. *Zinc and Copper Working Standards:* Dilute 5 and 10 mL of the Zinc and Copper Standards (see 6.10) to 1,000-mL with Extraction Reagent (see 6.3) corresponding to 0.4 Zn L^{-1} and 0.4 mg Cu L^{-1}. Following the manufacturer's guidelines, adjust the instrument to zero with Extraction Reagent (see 6.3). Atomize the 0.4 mg Zn, Cu L^{-1} standard and adjust the instrument reading to 100. Intermediate standards can be prepared to check for linearity. Higher concentrations can be used on instruments equipped with a 3-standard microprocessor.

6.12. *Procedure and Calculations:* With the 0.4 mg Zn, Cu L^{-1} standard (see 6.7), set the instrument reading to 100. Soil extracts can be read directly with appropriate dilutions when the instrument reading exceeds 100. Instrument reading × 0.4 = mg Zn, Cu dm^{-3} of soil. Alternately, instrument reading × 0.008 = kg Zn, Cu ha^{-1}. To convert kg Zn, Cu ha^{-1} to lbs A^{-1}, multiply by 0.891. These calculations are based on a volume of soil to a depth of 20 cm. For the rationale, see Mehlich (1972, 1973).

7. Decontamination Solution

7.1. *Decontamination Solution* (0.2% $AlCl_3 \cdot 6H_2O$): Dissolve 20 g aluminium chloride ($AlCl_3 \cdot 6H_2O$) in about 2,000 mL water and make to 10 liters with water. This solution volume can vary with the number of samples involved.

7.2. *Procedure:* Wash the extraction bottles (see 5.3), extraction vials (see 5.8), and funnels (see 5.7) with hot tap water, rinse with 0.2% $AlCl_3 \cdot 6H_2O$, then rinse with pure water. After placing filter paper into the funnels, rinse the paper with 0.2% $AlCl_3 \cdot 6H_2O$ followed with pure water. Allow to drain. All washing apparatus should be constructed from stainless steel or plastic.

8. Extraction Procedure, Analysis, and Calculations

8.1. *Extraction:* Scoop 5-cm³ air-dry < 10-mesh (2-mm) soil into an acid-washed 100-mL Extraction Bottle (see 5.3), add 50 mL Extraction Reagent (see 6.3), and shake for five minutes on a reciprocating shaker (see 5.4). Filter and collect the filtrate.

8.2. *Blank:* An Extraction Bottle (see 5.3) without soil but added Extracting Reagent (see 6.3) should be carried through the shaking and filtering procedure to serve as the *blank,* substracting the B, Cu, Mn, and Zn determined from the soil analysis result.

8.3. *Analysis:* Cu, Mn, and Zn in the filtrate can be determined by either AAS or ICP-AES, and B by UV-VIS spectrophotometry or ICP-AES (see 2.2). Since instruments and analysis procedures vary depending upon their operating conditions, no specific details are given.

8.4. *Calculations:* Follow the procedures given in Sections 6.6, 6.9, and 6.12.

9. Effects of Storage

9.1. Soils may be stored in an air-dried condition for several months with no effects on extractable B, Mn, Cu, and Zn content. The effect of time and temperature is discussed by Houba and Novozamsky (1998).

10. Interpretation

10.1. *Boron:* Critical B soil test levels are similar to that for hot water-extraction B (Shuman et al., 1992).

10.2. *Manganese:* Calibration of the Mn soil test with this extractant is based on extractable Mn and soil pH (Mascagni and Cox, 1983). Equations predicting the Mn availability index (MnAI) for soybeans and corn are as follows:

$$\text{Soybean MnAI} = 101.2 + 0.6 \text{ (MnI)} - 15.2 \text{ (pH)}$$

$$\text{Corn MnAI} = 108.2 + 0.6 \text{ (MnI)} - 15.2 \text{ (pH)}$$

The critical soil test MnAI = 4 mg Mn dm^{-3}, which is equal to a 25-soil test index. Because of the limited soil test calibration for other crops, calculation of the MnAI for these crops is based on their sensitivity to Mn, as compared to corn or soybeans.

For example, the soybean MnAI is used to predict Mn needs for small grains, since their sensitivity is closely related to that of soybeans.

10.3. *Copper:* Critical Cu soil test level was established with the Mehlich No. 3 extractant (Mehlich, 1984). The critical level is 0.5 mg Cu dm^{-3} which equates to a soil test index of 25.

10.4. *Zinc:* Critical Zn soil test level by this procedure is 1.0 mg Zn dm^{-3}, which equates to a soil test index of 25. A Zn availability index (ZnAI) has been established for mineral, mineral-organic, and organic soils and is based on the relationship between extractable Zn and soil pH (Junus, 1984). These values are as follows:

$$\text{ZnAI (mineral soils)} = \text{ZnI} \times 1.0.$$

$$\text{ZnAI (mineral-organic soils)} = \text{ZnI} \times 1.25.$$

$$\text{ZnAI (organic soils)} = \text{ZnI} \times 1.66.$$

8.4 AMMONIUM BICARBONATE–DTPA EXTRACTION

1. Principle of the Method

1.1. The Extraction Reagent is 1 M ammonium bicarbonate (NH$_4$HCO$_3$) in 0.005 M DTPA adjusted to a pH of 7.6 (Soltanpour and Schwab, 1977; Soltanpour and Workman, 1979; Soltanpour, 1991). The original pH of 7.6 allows DTPA to chelate and extract iron (Fe) and other metals.

1.2. This method is highly correlated with the DTPA (Lindsay and Norvell, 1979) method for zinc (Zn), iron (Fe), manganese (Mn), and copper (Cu) (Soltanpour and Schwab, 1977; Soltanpour et al., 1976). The method is described by Anon. (1994c).

2. Range, Sensitivity, and Methods of Analysis

2.1. The range and sensitivity are the same as those for the DTPA.

2.2. The sensitivity will vary with the type of instrument used, wavelength selected, and method of excitation.

2.3. The commonly used methods of analysis are flame emission, atomic absorption spectrophotometry (AAS) (Watson and Isaac, 1990; Wright and Stuczynski, 1996), or inductively coupled plasma emission spectrometry (ICP-AES) (Soltanpour, 1991; Soltanpour et al., 1996, 1998).

3. Interferences

3.1. The Extraction Reagent is unstable with regard to pH and must be kept under mineral oil to prevent a pH change. Ideally, it should be made as needed.

3.2. Use stainless steel soil sampling tubes and polyvinyl chloride mixing buckets for field soil sampling to prevent contamination with trace elements. Use high density aluminium oxide grinders equipped with stainless steel sieves to prevent soil contamination with trace elements. If the above grinder is not available, test other grinders with pure sand to make sure they do not contaminate the soil.

3.3. Grinding force and time and the amount of soil in the grinder should be adjusted so that extracted levels of trace elements are equivalent to those obtained with a wooden roller (Soltanpour et al., 1979). Kahn (1979) discusses the effect of soil particle size on the level of extractable Fe, Zn, and Cu.

4. Precision and Accuracy

4.1. A coefficient of variability ranging from 5 to 10% can be expected for DTPA test for micronutrients.

5. Apparatus

5.1. No. 10 (2-mm opening) sieve constructed from stainless steel or Nalgene.
5.2. Analytical balance.
5.3. 125-mL polyethylene conical flasks.
5.4. Eberbach reciprocating shaker (or equivalent), 180 oscillations per minute.
5.5. Whatman No. 42 filter paper (or equivalent), 12.5 cm.
5.6. Atomic absorption spectrophotometer (AAS) or inductively coupled plasma atomic emission spectrometer (ICP-AES).
5.7. A pH meter readable to 0.01 pH unit.
5.8. Funnel rack.
5.9. Accurate automatic diluter.
5.10. Volumetric flasks and pipettes as required for preparation of reagents, and standard solutions.

6. Reagents (use reagent grade chemical and pure water)

6.1. *Extraction Reagent* [Ammonium bicarbonate (AB)-DTPA]: 0.005 M DTPA (diethylenetriaminepentaacetic acid) solution is obtained by adding 9.85 g DTPA (acid form) to 4,500 mL water in a 5,000-mL volumetric flask. Shake for five hours constantly to dissolve the DTPA. Bring to 5,000 mL with water. This solution is stable with regard to pH. To 900 mL of the 0.005 M DTPA solution, add 79.06 g ammonium bicarbonate (NH_4HCO_3) gradually and stir gently with a rod to facilitate dissolution and to prevent effervescence when bicarbonate is added. The solution is diluted to 1,000 mL with the 0.005 M DTPA solution and mixed gently with a rod. The pH is adjusted to 7.6 with slow agitation with a rod by adding 2 M hydrochloric acid (HCl) solution. The AB-DTPA solution must be stored under mineral oil. Check the pH after storage and adjust it with a 2 M HCl solution dropwise, if necessary. The cumulative volume of HCl added should not exceed 1 mL L^{-1} limit, after which a fresh solution should be prepared.

7. Procedures

7.1. *Extraction Method:* Place 10 g air-dry soil <10-mesh (2-mm) in a 125-mL conical flask, add 20 mL Extraction Reagent (see 6.1), and shake on an Eberbach reciprocal shaker or an equivalent shaker for exactly 15 minutes at 180 cycles per minute with flasks kept open. The extracts are then filtered through Whatman 42 filter paper (Lindsay and Norvell, 1969).

7.2. Blank: A extraction bottle without soil but added Extracting Reagent (see 6.1) should be carried through the shaking and filtering procedure to serve as the *blank*, substacting the Cu, Fe, Mn, and Zn determined from the soil analysis result.

7.3. Analysis: Cu, Fe, Mn, and Zn in the filtrate can be determined by either AAS or ICP-AES (see 2.3). Since instruments vary in their operating conditions, no specific details are given.

8. Standards

8.1. Micronutrient Standards: Commercially prepared 1,000 mg L^{-1} standards, or weigh 1,000 mg each of Zn, Fe, Mn, and Cu metal separately, for AAS determination, into 250-mL volumetric flasks. Add 10 mL water and then 5 mL conc. nitric acid (HNO_3) into the flasks and warm gently to dissolve the metals. Then boil to expel the oxides of nitrogen. Bring the solution to dryness. Transfer into a 1,000-mL volumetric flask and bring to volume with Extraction Reagent (see 6.1) to obtain a primary standard of 1,000 mg element L^{-1}. Dilute the primary standard with Extraction Reagent (see 6.1) to bracket the concentrations found in soil extracts. For normal soils a concentration range of 0.1 to 10 mg L^{-1} can be used.

9. Calculations

9.1. Mg Cu, Fe, Mn, or Zn kg^{-1} in soil = mg Cu Fe, Mn, or Zn L^{-1} in extract × 2.

10. Effects of Storage

10.1. Air-drying and storage will not have any significant effect on the levels of micronutrients. The influence of time and temperature is discussed by Houba and Novozamsky (1998).

10.2. The Extraction Reagent (see 6.1) must be stored under mineral oil.

11. Interpretation

The following tables give an interpretation of the index values for Cu, Fe, Mn, and Zn. These are general guidelines and should be verified under different soil-climate-crop-management combinations.

Index values for Cu, Fe, Mn, and Zn

Category	Cu	Fe	Mn	Zn
		mg kg^{-1} in soil		
Low	0.0 – 0.2	0.0 – 3.0	0.0 – 0.5	0.0 – 0.9
Marginal	0.3 – 0.5	3.1 – 5.0	0.6 – 1.0	1.0 – 1.5
Adequate	>0.5	>5.0	>1.0	>1.5

12. Comments

12.1. For micronutrient tests soils should not be overground, and grinding should simulate the crushing force of a wooden roller (Soltanpour and Schwab, 1979).

Oven-drying of soils will change the test results and should be avoided (Soltanpour et al., 1976).

8.5 DTPA EXTRACTION

1. Principle of the method

1.1. The theoretical basis for the DTPA extraction is the equilibrium of the metal in the soil with the chelating agent. A pH level of 7.3 enables DTPA to extract iron (Fe) and other metals.

1.2. The use of DTPA as an extraction reagent was developed by Lindsay and Norvell (1978), and the method is described by Johnson (1992), Liang and Karamanos (1993), and Anon. (1994c).

2. Range, Sensitivity, and Methods of Analysis

2.1. Copper (Cu), iron (Fe), manganese (Mn), and zinc (Zn) can be extracted and determined in soil concentrations of 0.1 to 10 mg L^{-1} without dilution. The range and upper limits may be extended by diluting the extracting filtrate prior to analysis.

2.2. The sensitivity will vary with the type of instrument used and wavelength selected.

2.3. The determination of Cu, Fe, Mn, and Zn in the filtrate is most commonly done by either atomic absorption spectrophotometry (AAS) (Watson and Isaac, 1990; Wright and Stuczynski, 1996) or by inductively coupled plasma emission spectrometry (ICP-AES) (Soltanpour et al., 1996, 1998).

3. Interferences

3.1. Triethanolamine (TEA) is used to keep the pH close to 7.3.

3.2. All apparatus that will come in direct contact with the extractant and extraction filtrate of the soil must be thoroughly washed and rinsed in pure hydrochloric acid (HCl) and pure water before use. Avoid contact with rubber and metals.

3.3. Contamination of soil samples may occur in either the sampling equipment or soil-grinding equipment, especially for Zn and Fe.

3.4. The effect of wetting and drying on DTPA-extractable Fe, Zn, Mn, and Cu is discussed by Kahn and Soltanpour (1978).

4. Precision and Accuracy

4.1. Repeated analysis of the same soil with medium concentration ranges of Cu, Fe, Mn, and Zn will give coefficients of variability of from 10 to 15%. A major portion of the variance is related to the heterogeneity of the soil rather than to the extraction process or method of analysis.

5. Apparatus

5.1. No. 10 (2-mm opening) sieve.
5.2. 8.5-cm^3 scoop, volumetric.

5.3. Extraction flask, 125-mL polyethylene conical flasks.
5.4. Mechanical reciprocating shaker, 180 oscillations per minute.
5.5. Filter funnel, 11 cm.
5.6. Whatman No. 42 ashless filter paper (or equivalent), 12.5 cm.
5.7. Atomic absorption spectrophotometer (AAS) or inductively coupled plasma emission spectrometer (ICP-AES).
5.8. pH meter with reproducibility to at least 0.1 pH unit and glass electrode paired with a calomel reference electrode.
5.9. Analytical balance.
5.10. Volumetric flasks, pipettes, and microburet as required for preparation of reagents and standard solution.

6. Reagents (use reagent grade chemicals and pure water)

6.1. *Extraction Reagent* (DTPA-diethylenetriaminepentaacetic acid): Weigh 1.96 DPTA {[(HOCOCH$_2$)$_2$NCH$_2$CH$_2$]$_2$NCH$_2$COOH} into a 1,000-mL volumetric flask. Add 14.92 g triethanolamine (TEA) and bring to volume to approximately 950mL with water. Add 1.47 g calcium chloride (CaCl$_2$·2H$_2$O) and bring to 1,000-mL with water while adjusting the pH to exactly 7.3 with 6N hydrochloric acid (HCl). The final concentration will be 0.005 M DTPA, 0.1 M TEA, and 0.01 M CaCl$_2$·2H$_2$O. (*Note:* The DTPA reagent should be the acid form.)

6.2. *Zinc Standard* (1,000 mg Zn L^{-1}): Commercially prepared standard, or weigh 1.000 g pure Zn metal into 5–10 mL conc. hydrochloric acid (HCl). Evaporate almost to dryness and dilute to 1,000 mL with Extraction Reagent (see 6.1). Prepare working standards by diluting aliquots of the stock solution standard with Extraction Reagent (see 6.1) to cover the anticipated range in the concentration to be found in the soil extraction filtrate. Working standards from 0.1 to 10 mg Zn L^{-1} should be sufficient for most soils.

6.3. *Iron Standard* (1,000 mg Fe L^{-1}): Commercially prepared standard, or weigh 1.000 g pure Fe wire into 5–10 mL conc. hydrochloric acid (HCl). Evaporate almost to dryness and dilute to 1,000 mL with Extraction Reagent (see 6.1). Prepare working standards by diluting aliquots of the stock solution standard with Extraction Reagent (see 6.1) to cover the anticipated range in concentration to be found in the soil extraction filtrate. Working standards from 0.1 to 10 mg Fe L^{-1} should be sufficient for most soils.

6.4. *Manganese Standard* (1,000 mg Mn L^{-1}): Commercially prepared standard, or weigh 1.582 g manganese oxide (MnO$_2$) into 5 mL conc. hydrochloric acid (HCl). Evaporate almost to dryness and dilute to 1,000 mL with Extraction Reagent (see 6.1). Prepare working standards by diluting aliquots of the stock solution standard with the Extraction Reagent (see 6.1) to cover the anticipated range in concentration to be found in the soil extraction filtrate. Working standards from 0.1 to 10 mg Mn L^{-1} should be sufficient for most soils.

6.5. *Copper Standard* (1,000 mg Cu L^{-1}): Commercially prepared standard or weigh 1.000 g pure Cu metal and dissolve in a minimum amount conc. nitric acid (HNO$_3$) and add 5 mL conc. hydrochloric acid (HCl). Evaporate almost to dryness and dilute to 1,000 mL with Extraction Reagent (see 6.1). Prepare working standards by diluting aliquots of the stock solution with Extraction Reagent (see 6.1) to cover the anticipated range in concentration filtrate. Working standards from 0.1 to 10 mg Cu L^{-1} should be sufficient for most soils.

7. Procedure

7.1 *Extraction*: Weigh 10 g or measure 8.5-cm^3 (see 5.2) of air-dry <10-mesh (2-mm) soil into a 125-mL Extraction Flask (see 5.3), add 20 mL Extraction Reagent (see 6.1), and shake on a reciprocating shaker for two hours. Samples shaken longer than two hours will give high results because a final equilibrium of the metal and soil is not reached in two hours. Filter and collect the filtrate.

7.2. *Blank*: A extraction bottle without soil but added Extracting Reagent (see 6.1) should be carried through the shaking and filtering procedure to serve as the *blank*, substracting the Cu, Fe, Mn, and Zn determined from the soil analysis result.

7.3. *Analysis*: Cu, Fe, Mn, and Zn in the filtrate can be determined by either AAS or ICP-AES (see 2.3). Since instruments vary in their operating conditions, no specific details are given, recommend following the instrument manufacturer's instructions.

8. Calibration and Standards

8.1. *Working Standards*: Working standards are prepared as described in Sections 6.2 to 6.5. If element concentrations are found to be outside the range of the instrument or standards, suitable dilutions should be prepared, starting with a 1:2 soil extract to Extraction Reagent (see 6.1) dilutions.

8.2. *Calibration*: Procedures vary with instrument techniques and type of instrument. Every precaution should be taken and the manufacturer's recommendations should be followed in the operation and calibration of the instrument.

9. Calculation

9.1. The results are reported as kg ha^{-1} for a 20 cm depth: kg element ha^{-1} = mg element L^{-1} in extraction filtrate × 4.

If the extraction filtrate is diluted, the dilution factor must be applied. For expressing the results in mg kg^{-1} of soil, use the following formula: mg kg^{-1} in soil = mg L^{-1} in solution × 2.

10. Effects of Storage

10.1. Soils may be stored in an air-dry condition for several months with no effects on the amount of Cu, Fe, Mn, and Zn extracted. The influence of time and temperature is discussed by Houba and Novozamsky (1998).

11. Interpretation

11.1 An evaluation of the analysis results as well as accurate fertilizer recommendations for Cu, Fe, Mn, and Zn must be based on field response for each crop and local field conditions. Interpretative data for critical levels as established by Viets and Lindsay (1973) for Colorado soil are available. Boawn (1973) did work with DTPA for Zn on Washington soil.

12. Comments

12.1 Grinding can change the amount of DTPA-extractable micronutrients, especially Fe. Therefore, it is imperative that grinding procedures be standardized along with extraction procedures. Grinding should be equivalent to using a wooden roller to crush the soil aggregates (Soltanpour et al., 1976).

8.6 REFERENCES

Adriano, D. C. 1986. *Trace Elements in the Environment.* New York: Springer-Verlag.
Adriano, D. C. 1992. *Geochemistry of Trace Elements.* New York: Springer-Verlag.
Alley, M. M., D. C. Martens, M. G. Schnappinger, and G. W. Hawkins. 1972. Field calibration of soil tests for available zinc. *Soil Sci. Soc. Amer. Proc.* 36:621–624.
Anon. 1994a. Extractable phosphorus, potassium, calcium, magnesium, sodium, iron, manganese, and copper: Mehlich-3 method, In: *Plant, Soils, and Water Reference Methods for the Western Region.* Western Regional Extension Publication WREP 125, R. G. Gavlak, D. A. Horneck, and R. O. Miller (eds.). AK: University of Alaska, Fairbanks, 25–26.
Anon. 1994b. Extractable phosphorus, potassium, sulfate-sulfur, zinc, iron, manganese, copper, nitrate: Ammonium bicarbonate-DTPA method, In: *Plant, Soils, and Water Reference Methods for the Western Region.* Western Regional Extension Publication WREP 125, R. G. Gavlak, D. A. Horneck, and R. O. Miller (eds.), Fairbanks, AK: University of Alaska.
Anon. 1994c. Extractable zinc, copper, iron, and manganese: DTPA method, In: *Plant, Soils, and Water Reference Methods for the Western Region.* Western Regional Extension Publication WREP 125, R. G. Gavlak, D. A. Horneck, and R. O. Miller (eds.). Fairbanks, AK: University of Alaska, 29–30.
Anon. 1994d. Hot-water extractable boron: Azomethine-H method, In: *Plant, Soils, and Water Reference Methods for the Western Region.* Western Regional Extension Publication WREP 125. R. G. Gavlak, D. A. Horneck, and R.O. Miller (eds.). Fairbanks, AK: University of Alaska, 33–34.
Anon. 1996. The Micronutrients. *Soil Fertility Manual* Chapter 7, Norcross, GA: Potash and Phosphate Institute, 7:1–10.
Armour, J. D. and R. F. Brennan. 1999. Zinc, In: *Soil Analysis: An Interpretation Manual,* K. I. Peverill, L.A. Sparrow, and D. J. Reuter (eds.). Collingwood, Australia: CSIRO Publishing, 281–285.
Bell, R. W. 1999. Boron, In: 1999. *Soil Analysis: An Interpretation Manual,* K. I. Peverill, L. A. Sparrow, and D. J. Reuter (eds.). Collingwood, Australia: CSIRO Publishing, 309–317.
Berger, K. C. and E. Truog. 1940. Boron deficiency as revealed by plant and soil tests. *J. Amer. Soc. Agron.* 32:297–301.
Berger, K. C. and E. Truog. 1944. Boron tests and determination for soils and plants. *Soil Sci.* 57:25–36.
Boawn, L. C. 1971. Evaluation of DTPA-extractable zinc as a zinc soil test for Washington soils, In: *Proceedings of the 22nd Annual Pacific Northwest Fertilizer Conference.* Bozeman, MT. Pacific NW Plant Food Association, 143–147.
Bingham, F. T. 1982. Boron, In: *Methods of Soil Analysis, Part 2, Chemical and Microbiological Properties,* 2nd ed., A. L. Page (ed.). Madison, WI: American Society of Agronomy, 431–447.
Brennan, R. F. and E. Best. 1999. Copper, In: *Soil Analysis: An Interpretation Manual,* K. I. Peverill, L. A. Sparrow, and D. J. Renter (eds.). Collingwood, Australia: CSIRO Publishing.

Cox, F. R. and J. I. Wear (eds.). 1977. *Diagnosis and Correlation of Zinc Problems in Corn and Rice Production.* Southern Cooperative Series Bulletin 222. Raleigh, NC: North Carolina State Experiment Station.

Dahnke, W. C. and R. A. Olson. 1990. Soil test correlation, calibration, and recommendations, In: *Soil Testing and Plant Analysis,* 3rd ed. SSSA Book Series No. 3. R. L. Westerman (ed.). Madison, WI: Soil Science Society of America, 45–71.

Epstein, E. 1972. *Mineral Nutrition of Plants: Principles and Perspectives.* New York: John Wiley & Sons.

Gambrell, R. P. 1996. Manganese, In: *Methods of Soils Analysis, Part 3, Chemical Methods,* SSSA Book Series No. 5. D. L. Sparks (ed.) Madison, WI: Soil Science Society of America, 665–683.

Glass, D. M. 1989. *Plant Nutrition: An Introduction to Current Concepts.* Boston, MA: Jones and Bartlett Publishers.

Gupta, U. C. 1967. A simplified method for determining hot water-soluble boron in podzol soils. *Soil Sci.* 103:424–428.

Gupta, U. C. 1990. Boron, molybdenum, and selenium, In: *Soil Sampling and Methods of Analysis,* M. R. Carter (ed.). Boca Raton, FL: Lewis Publishers, 91–99.

Houba, Y. J. G. and I. Novozamsky. 1998. Influence of storage time and temperature of air-dried soils and extractable nutrients using 0.01 M $CaCl_2$. *Fresenicus J. Anal. Chem.* 360:362–365.

Isaac, R. A. 1992. Determination of boron by hot water extraction, In: *Reference Soil and Media Diagnostic Procedures for the Southern Region of the United States.* Southern Cooperative Series Bulletin No. 374, S. J. Donohue (ed.). Blacksburg, VA: Virginia Agricultural Experiment Station, 23–24.

John, M. K. 1973. A batch-handling technique for hot-water extraction of boron from soils. *Soil Sci. Soc. Amer. Proc.* 37:332–333.

Johnson, G. V. 1992. Determination of zinc, manganese, copper, and iron by DTPA extraction, In: *Reference Soil and Media Diagnostic Procedures for the Southern Region of the United States.* Southern Cooperative Series Bulletin No. 374. S. J. Donohue (ed.). Blacksburg, VA: Virginia Agricultural Experiment Station, 16–18.

Johnson, G. V. and P. E. Fixen. 1990. Testing soils for sulfur, boron, molybdenum, and chlorine, In: *Soil Testing and Plant Analysis,* 3rd ed., SSSA Book No. 3. R. L. Westerman (ed.). Madison, WI: Soil Science Society of America, 265–273.

Jones, J. B., Jr. 1998a. *The Plant Nutrition Manual.* Boca Raton, FL: CRC Press.

Jones, J. B., Jr. 1998b. Soil test methods: Past, present, and future. *Commun. Soil Sci. Plant Anal.* 29(11–14): 1543–1552.

Kabata-Pendias, A. and D. C. Adriano. 1995. Trace elements, In: *Soil Amendments and Environmental Quality,* J. E. Rechcigl (ed.). Boca Raton, FL: Lewis Publishers, 139–167.

Kabata-Pendias, A. and H. Pendias. 1995. *Trace Elements in Soils and Plants,* revised ed., New York: John Wiley & Sons.

Kahn, A. 1979. Distribution of DTPA-extractable Fe, Zn, and Cu in soil particle-size fractions. *Commun. Soil Sci. Plant Anal.* 10:1211–1218.

Kahn, A. and P. N. Soltanpour. 1978. Effect of wetting and drying on DTPA-extractable Fe, Zn, Mn, and Cu in soils. *Commun. Soil Sci. Plant Anal.* 9:193–202.

Keren, R. 1996. Boron, In: *Methods of Soils Analysis, Part 3, Chemical Methods.* SSSA Book Series No. 5. D. L. Sparks (ed.), Madison, WI: Soil Science Society of America, 603–626.

Liang, J. and R. E. Karamanos. 1993. DTPA-extractable Fe, Mn, Cu and Zn, In: *Soil Sampling and Methods of Analysis,* M. R. Carter (ed.). Boca Raton, FL: Lewis Publishers, 87–90.

Lindsay, W. L. 1991. Inorganic equilibria affecting micronutrients in soils, In: *Micronutrients in Agriculture*, SSSA Book Series No. 4, J. J. Mortvedt (ed.). Madison, WI: Soil Science Society of America, 89–112.

Lindsay, W. L. and W. A. Norvell. 1978. Development of a DTPA micronutrient soil test for zinc, iron, manganese, and copper. *Soil Sci. Soc. Amer. J.* 42:421–428.

Loeppert, R. L. and W. P. Inskeep. 1996. Iron, In: *Methods of Soils Analysis, Part 3, Chemical Methods*, SSSA Book Series No. 5, D. L. Sparks (ed.) Madison, WI: Soil Science Society of America.

Martens, D. C. and W.L. Lindsay. 1990. Testing soils for copper, iron, manganese, and zinc, In: *Soil Testing and Plant Analysis*, 3rd ed. SSSA Book Series No. 3, R. L. Westerman (ed.). Madison, WI: Soil Science Society of America, 229–273.

Martens, D. C. and D. T. Westerman. 1991. Fertilizer applications for correcting micronutrient deficiencies, In: *Micronutrients in Agriculture*, 2nd ed., SSSA Book Series No. 4. J.J. Mortvedt (ed.). Madison, WI: Soil Science Society of America, 549–592.

Mascagni, H. J. and F. R. Cox. 1985. Calibration of a manganese availability index for soybeans soil test data. *Soil Sci. Soc. Amer. J.* 49:382–386.

Makarim, A. K. and F. R. Cox. 1983. Evaluation of the need for copper with several soil extractants. *Agron. J.* 75:493–496.

McFarlane, J. D. 1999. Iron, In: *Soil Analysis: An Interpretation Manual*, K. I. Peverill, L. A. Sparrow, and D. J. Reuter (eds.). Collingwood, Australia: CSIRO Publishing, 295–301.

Mehlich, A. 1953. *Determination of P, Ca, Mg, K, Na and NH_4* (mimeo). Raleigh, NC: North Carolina Soil Testing Division.

Mehlich, A. 1972. Uniformity of expressing soil test results. A case for calculating results on a volume basis. *Commun. Soil Sci. Plant Anal.* 3:417–424.

Mehlich, A. 1973. Uniformity of soil test results as influenced by volume weight. *Commun. Soil Sci. Plant Anal.* 4:475–486.

Mehlich, A. 1978. New extractant for soil test evaluation of phosphorus, potassium, calcium, magnesium, sodium, manganese and zinc. *Commun. Soil Sci. Plant Anal.* 9:477–492.

Mehlich, A. 1984. Mehlich 3 soil test extractant: A modification of Mehlich 2 extractant. *Commun. Soil Sci. Plant Anal.* 15:1409–1416.

Maynard, D. N. and G. J. Hochmuth (eds.). 1997. *Knott's Handbook for Vegetable Growers*, 4th ed. New York: John Wiley & Sons.

Mikkelsen, R. L. and J. J. Camberato. 1995. Potassium, sulfur, lime, and micronutrient fertilizers, In: *Soil Amendments and Environmental Quality*, J. E. Rechcigl (ed.). Boca Raton, FL: Lewis Publishers, 109–137.

Moraghan, J. T. and H. J. Mascagni, Jr. 1991. Environmental and soil factors affecting micronutrient deficiencies and toxicities, In: *Micronutrients in Agriculture*, 2nd ed., SSSA Book Series No. 4. J. J. Mortvedt (ed.). Madison, WI: Soil Science Society of America, 371–425.

Nelson, J. L., L. C. Boawn, and F. G. Viets, Jr. 1959. A method for assessing zinc status of soils using acid extractable zinc and "titratable alkalinity" values. *Soil Sci.* 88:275–283.

Odom, J. W. 1980. Kinetics of the hot-water soluble boron soil test. *Commun. Soil Sci. Plant Anal.* 11:759–769.

Pais, I. and J. B. Jones, Jr. 1998. *The Handbook of Trace Elements*. Boca Raton, FL: St. Lucie Press.

Peck, T. R., J. T. Cope, Jr., and D. A. Whitney (eds.). 1977. *Soil Testing: Correlating and Interpreting the Analytical Results*, ASA Special Publication No. 29. Madison, WI: American Society of Agronomy.

Perkins, H. F. 1970. A rapid method of evaluating the zinc status of coastal plain soils. *Commun. Soil Sci. Plant Anal.* 1:35–42.

Reed, S. T. and D. C. Martens. 1996. Copper and zinc, In: *Methods of Soils Analysis, Part 3, Chemical Methods,* SSSA Book Series No. 5. D. L. Sparks (ed.). Madison, WI: Soil Science Society of America, 703–722.

Rechcigl, J. E., G. G. Payne, and C. A. Sanchez. 1992. Comparison of various soil drying techniques on extractable nutrients. *Commun. Soil Sci. Plant Anal.* 23:2347–2363.

Reisenauer, H. M., L. M. Walsh, and R. G. Hoeft. 1973. Testing soils for sulfur, boron, molybdenum, and chlorine, In: *Soil Testing and Plant Analysis, Revised Edition,* L. M. Walsh and J. D. Beaton (eds.). Madison, WI: Soil Science Society of America, 173–200.

Ross, G. J. and C. Wang. 1990. Extractable Al, Fe, Mn, and Si, In: *Soil Sampling and Methods of Analysis,* M. R. Carter (ed.). Boca Raton, FL: Lewis Publishers, 239–246.

Schmisek, M. E., L. J. Cihacek, and L. J. Swenson. 1989. Relationships between the Mehlich-III soil test extraction procedure and standard soil test methods in North Dakota. *Commun. Soil Sci. Plant Anal.* 29:1719–1729.

Shuman, L. M., V. A. Bandel, S. J. Donohue, R. A. Isaac, R. M. Lippert, J. T. Sims, and M. R. Tucker. 1992. Comparison of Mehlich-1 and Mehlich-3 extractable soil boron with hotwater extractable boron. *Commun. Soil Sci. Plant Anal.* 23:1–14.

Sims, J. T. 1989. Comparison of Mehlich 1 and Mehlich 3 extractants for P, K, Ca, Mg, Mn, Cu, and Zn in Atlantic coastal plain soils. *Commun. Soil Sci. Plant Anal.* 20:1707–1726.

Sims, J. T. 1995. Recommended soil tests for the micronutrients: Manganese, zinc, copper, and boron, In: *Recommended Soil Testing Procedures for the Northeastern United States.* Northeast Regional Publication No. 493 (revised), J. T. Sims and A. M. Wolf (eds.). Newark, DE: University of Delaware, 40–45.

Sims, J. T. and G. V. Johnson. 1991. Micronutrient soil tests, In: *Micronutrients in Agriculture,* SSSA Book Series No. 4. J. J. Mortvedt (ed.). Madison, WI: Soil Science Society of America.

Soltanpour, P. N. 1991. Determination of nutrient availability and elemental toxicity by AB-DTPA soil test and ICPS. *Adv. Soil Sci.* 16:165–190.

Soltanpour, P. N. and A. P. Schwab. 1977. A new soil test for simultaneous extraction of macro- and micronutrients in alkaline soils. *Commun. Soil Sci. Plant Anal.* 8:195–207.

Soltanpour, P. N., and S. M. Workman. 1979. Modification of NH_4HCO_3-DTPA soil test to omit carbon black. *Commun. Soil Sci. Plant Anal.* 10:1411–1420.

Soltanpour, P. N., A. Khan, and W. L. Lindsay. 1976. Factors affecting DTPA-extractable Zn, Fe, Mn and Cu. *Commun. Soil Sci. Plant Anal.* 7:797–821.

Soltanpour, P. N., A. Khan, and A. P. Schwab. 1979. Effect of grinding variables on the NH_4HCO_3-DTPA soil test values for Fe, Zn, Mn, Cu, P, and K. *Commun. Soil Sci. Plant Anal.* 10:903–909.

Soltanpour, P. N., C. W. Johnson, S. M. Workman, J. B. Jones, Jr., and R. O. Miller. 1996. Inductively coupled plasma emission spectrometry and inductively coupled plasma-mass spectroscopy, In: *Methods of Soil Analysis, Part 3, Chemical Method,* SSSA Book Series No. 5. D. L. Sparks (ed.). Madison, WI: Soil Science Society of America, 91–139.

Soltanpour, P. N., C. W. Johnson, S. M. Workman, J. B. Jones, Jr., and R. O. Miller. 1998. Advances in ICP emission and ICP mass spectrometry. *Adv. Agron.* 64:28–113.

Stevenson, F. J. 1991. Organic matter—Micronutrient reactions in soil, In: *Micronutrients in Agriculture,* 2nd ed., SSSA No. 4. J. J. Mortvedt (ed.). Madison, WI: Soil Science Society of America, 145–186.

Tucker, M. R. 1992. Determination of zinc, manganese, and copper by Mehlich 3 extraction, In: *Reference Soil and Media Diagnostic Procedures for the Southern Region of the*

United States. Southern Cooperative Series Bulletin No. 374, S. J. Donohue (ed.). Blacksburg, VA: Virginia Agricultural Experiment Station, 19–22.

Uren, N. C. 1999. Manganese, In: *Soil Analysis: An Interpretation Manual*, K. I. Peverill, L. A. Sparrow, and D. J. Reuter (eds.). Collingwood, Australia: CSIRO Publishing, 287–294.

Viets, Jr., F. G. and L. C. Boawn, 1965. Zinc, In: *Methods of Soil Analysis, Part 2*, Agronomy No. 9, C. A. Black (ed.). Madison, WI: American Society of Agronomy, 1090–1101.

Viets, Jr., F. G. and W.L. Lindsay. 1973. Testing soils for zinc, copper, manganese, and iron, In: *Soil Testing and Plant Analysis, revised edition.* L. M. Walsh and J. D. Beaton (eds.). Madison, WI: Soil Science Society of America, 153–172.

Watson, M. E. 1998. Boron test, In: *Recommended Chemical Soil Test Procedures for the North Central Region,* North Central Regional Publication No. 221 (revised). J. R. Brown (ed.). Columbia, MO: Missouri Agricultural Experiment Station SB 1001, University of Missouri, 45–48.

Watson, M. E. and R. A. Isaac. 1990. Analytical instrumentation for soil and plant analysis, In: *Soil Testing and Plant Analysis,* 3rd ed., SSSA Book Series No. 3. R. L. Westerman (ed.). Madison, WI: Soil Science Society of America, 691–740.

Wear, J. I. and C. E. Evans. 1968. Relationships of zinc uptake of corn and sorghum to soil zinc measured by three extractants. *Soil Sci. Soc. Amer. Proc.* 32:543–546.

Wear, J. I. 1965. Boron, In: *Methods of Soil Analysis, Part 2,* Agronomy No. 9, C. A. Black (ed.). Madison, WI: American Society of Agronomy, 1059–1063.

Wear, J. I. and C. E. Evans. 1968. Relationship of zinc uptake by corn and sorghum to soil zinc measured by three extractants. *Soil Sci. Soc. Amer. Proc.* 32:543–546.

Whitney, D. A. 1998. Micronutrients: Zinc, iron, manganese, and copper, In: *Recommended Chemical Soil Test Procedures for the North Central Region,* North Central Regional Research Publication No. 221 (revised), J. R. Brown (ed.). Columbia, MO: Missouri Agricultural Experiment Station SB 1001, University of Missouri, 41–44.

Wolf, B. 1971. The determination of boron in soil extracts, plant materials, composts, manures, water, and nutrient solutions. *Commun. Soil. Sci. Plant Anal.* 2:363–374.

Wolf, B. 1974. Improvements in the Azomethine-H method for the determination of boron. *Commun. Soil Soc. Plant Anal.* 5:39–44.

Wright, R. J. and T. I. Stuczynski. 1996. Atomic absorption and flame emission spectrometry, In: *Methods of Soil Analysis, Part 3, Chemical Method,* SSSA Book Series No. 5, D. L. Sparks (ed.), Madison, WI: Soil Science Society of America, 65–90.

9 Heavy Metals

The heavy metals have been variously classified, one being those elements that have atomic weights greater than 55. Under this classification, the micronutrients copper (Cu, atomic weight 63.54), iron (Fe, atomic weight 55.85), manganese (Mn, atomic weight 54.993), molybdenum (Mo, atomic weight 95.95), and zinc (Zn, atomic weight 65.38) would be so identified, elements that are covered in Chapter 8. Elements considered in this chapter as toxic to plants and/or animals (Adriano, 1986; Risser and Baker, 1990; Kabata-Pendias and Pendias, 1994; Pais and Jones, 1997) are cadmium (Cd), chromium (Cr), mercury (Hg), nickel (Ni), and lead (Pb); these are the elements covered in this chapter as the heavy metals. Three extraction methods are described.

The total heavy metal content of agricultural soils in the United States was determined by Holmgren et al. (1993), with a summary table given in the Potash & Phosphate Institute's *Bulletin 1998–2* (Anon, 1998). Based on the assay of 2,771 soils, the means and range in contents (in mg kg^{-1}) were: Cd-0.155 (0.037-0.531); Zn-41.1 (11.4-91.2); Cu-15.5 (5.3-64.8); Ni-17.1 (6.2-50.5); and Pb-104. (6.7-19.8). The heavy metal content of soils and plants and its effect on plant growth and development may be found in the books by Adriano (1986), Kabata-Pendis and Pendias (1994), and Pais and Jones (1997), and the article by Kabata-Pendias and Adriano (1995). Testing soils for the heavy metals has been discussed by Risser and Baker (1990), Amacher (1996), Bartlett and James (1996), Crock (1996), and Radojevic and Bashkin (1999), and waste-amended soil by Johnson and Donohue (1992).

9.1 AMMONIUM BICARBONATE–DTPA EXTRACTION

1. Principle of the Method

1.1. The Extraction Reagent is 1 M ammonium bicarbonate (NH$_4$HCO$_3$) in 0.005 M DTPA (AB-DTPA) adjusted to a pH of 7.6 (Soltanpour and Schwab, 1977; Soltanpour and Workman, 1979; Soltanpour, 1991). The original pH of 7.6 allows DTPA to chelate and extract iron (Fe) and other metals.

1.2. This method can be used for determination of bioavailability and biotoxicity of boron (B), molybdenum (Mo), nickel (Ni), arsenic (As), cadmium (Cd), lead (Pb), and selenium (Se) in addition to the above elements (Amacher, 1990; Baker and Amacher, 1982; Bartlett and James, 1996; Huang and Fuji, 1990; Riser and Baker, 1990; Soon and Abboud, 1993) in manure-amended, sludge-amended, mine spoil-contaminated and nonamended soils (Rappaport et al., 1987, 1988; Soltanpour, 1991; Johnson and Donohue, 1992).

2. Range, Sensitivity, and Methods of Analysis

2.1. The range and sensitivity are the same as those for the DTPA for micronutrients (Soltanpour et al. 1976).

2.2. The sensitivity will vary with the type of instrument used and the wavelength selected.

2.3. Commonly used elemental assay methods are either atomic absorption spectrophotometry (AAS) (Watson and Isaac, 1990; Wright and Stuczynski, 1996) or inductively coupled plasma emission spectrometry (Soltanpour, 1991; Soltanpour et al. 1996, 1998).

3. Interferences

3.1. The pH of AB-DTPA Extraction Reagent is unstable and should be kept under mineral oil to prevent a pH change. Ideally, it should be made as needed.

3.2. Use stainless steel soil sampling tubes and polyvinyl chloride mixing buckets for field soil sampling to prevent contamination with trace elements. Use high density aluminum oxide grinders equipped with stainless steel sieves to prevent soil contamination with trace elements. If the above grinder is not available, test other grinders with pure sand to make sure they do not contaminate the soil.

3.3. Grinding force and time and the amount of soil in the grinder should be adjusted so that extracted levels of trace elements are equivalent to those obtained with a wooden roller (Soltanpour et al., 1979).

3.4. The distribution of Fe, Zn, and Cu in soil particle-sized fractions is discussed by Kahn (1979) and grinding by Soltanpour et al. (1979).

4. Precision and Accuracy

4.1. A coefficient of variability ranging from 5 to 10% can be expected for different determinations. Accuracy is comparable DTPA test for the micronutrients (Lindsay and Norvell (1978).

5. Apparatus

5.1. No. 10 (2-mm opening) sieve constructed from stainless steel or Nalgene.
5.2. Analytical balance.
5.3. 125-mL polyethylene conical flasks.
5.4. Eberbach reciprocating shaker (or equivalent), 180 oscillations per minute.
5.5. Whatman No. 42 filter paper (or equivalent), 12.5 cm.
5.6. Atomic absorption spectrophotometer (AAS) or inductively coupled plasma emission spectrometer (ICP-AES).
5.7. pH meter readable to 0.01 pH units.
5.8. Funnels and funnel rack.
5.9. Accurate automatic diluter.
5.10. Volumetric flasks and pipettes as required for preparation of reagents and standard solutions.

6. Reagents (use reagent grade chemicals and pure water)

6.1. *Extraction Reagent* [Ammonium bicarbonate (AB)-DTPA]: 0.005 M DTPA (diethylenetriaminepentaacetic acid) solution is obtained by adding 9.85 g DTPA (acid form) to 4,500 mL water in a 5,000-mL volumetric flask. Shake for five hours constantly to dissolve the DTPA. Bring to 5,000 mL with water. This solution is stable with regard to pH. To 900 mL of the 0.005 M DTPA solution, add 79.06 g ammonium bicarbonate (NH_4HCO_3) gradually and stir gently with a rod to facilitate dissolution and to prevent effervescence when bicarbonate is added. The solution is diluted to 1,000 mL with the 0.005 M DTPA solution and mixed gently with a rod. The pH is adjusted to 7.6 with slow agitation with a rod by adding 2 M hydrochloric acid (HCl). The AB-DTPA solution must be stored under mineral oil. Check the pH after storage and adjust it with a 2 M HCl solution dropwise, if necessary. The cumulative volume of HCl added should not exceed 1 mL L^{-1} limit, after which a fresh solution should be prepared.

7. Procedures

7.1. *Extraction Method:* Weigh 10 g air-dry soil <10-mesh (2-mm) into a 125-mL conical flask, add 20 mL Extraction Reagent (see 6.1), and shake on an Eberbach reciprocal shaker or an equivalent shaker for exactly 15 minutes at 180 cycles per minute with flasks kept open. The extracts are then filtered through Whatman 42 filter paper (Soltanpour and Workman, 1979).

7.2. *Analysis:* Heavy metals in the filtrate are determined either by AAS or ICP-AES (see 2.3). Since instruments vary in their operating conditions, no specific details are given. The standard solutions of these ions are made in Extraction Reagent (see 6.1).

8. Standards

8.1. *Elemental Standards* (1,000 mg L^{-1}): Commercially prepared standards, or weigh 1,000 mg of each element separately, for atomic absorption spectrophotometry, into 250-mL volumetric flasks. Introduce 10 mL water and then 5 mL conc. nitric acid (HNO_3) into the flasks warm gently to dissolve the metals. Then boil to expel the oxides of N. Bring the solution to dryness. Transfer into a 1,000-mL volumetric flask and bring to volume with Extraction Reagent (see 6.1) to get a primary standard of 1,000 mg L^{-1}. Dilute the primary standard with Extraction Reagent (see 6.1) to bracket the concentrations found in soil extracts. For normal soils a concentration range of 0.1 to 10 mg L^{-1} can be used.

9. Calculations

9.1. mg element kg^{-1} in soil = mg element L^{-1} in extract $\times$ 2.

10. Effects of Storage

10.1. Air-drying and storage will not have any significant effect on the levels of the elements.

10.2. The Extraction Reagent (see 6.1) must be stored under mineral oil.

11. Interpretation

11.1. The heavy metal content of soils and plants and their effect on plant growth and development may be found in the books by Adriano (1986), Kabata-Pendias and Pendias (1994), and Pais and Jones (1997), and the article by Kabata-Pendias and Adriano (1995).

12. Comments

12.1. For micronutrient tests soils should not be over-ground and grinding should simulate the crushing force of a wooden roller (Soltanpour et al., 1979). Separation by particle size can affect the level of extraction (Kahn, 1979). Oven-drying of soils will change the test results and should be avoided (Soltanpour et al., 1976). These same conditions may occur with the other heavy metals.

9.2 DTPA EXTRACTION

1. Principle of the Method

1.1. The DTPA chelation procedure offers a favorable combination of stability constants for the simultaneous complexing of cadmium (Cd), copper (Cu), and nickel (Ni), and zinc (Zn) (Lindsay, 1979). The theoretical basis for the DTPA extraction is the equilibrium of the metal in the soil with the chelating agent. The 7.3 pH, which is buffered with triethanolamine (TEA), prevents excess dissolution of the trace (heavy) metals.

1.2. The use of DTPA as an extracting agent was developed by Lindsay and Norvell (1978).

2. Range, Sensitivity, and Methods of Analysis

2.1. Concentration ranges for each element depend on the instrument selected for metal determination. With atomic absorption spectrometry, Cd, Cu, Ni, and Zn can be determined in soil concentrations of 0.1 to 10 mg kg^{-1}. With inductively coupled plasma emission spectrometry (ICP-AES), the linear range is up to several orders of magnitude greater, depending on the element being determined. Range and upper limits may be extended by diluting the extracting filtrate prior to analysis.

2.2. The sensitivity will vary with the type of instrument used and the wavelength selected.

2.3. Commonly used elemental assay methods are either atomic absorption spectrophotometry (AAS) (Watson and Isaac, 1990; Wright and Stuczynski, 1996) or inductively coupled plasma emission spectrometry (Soltanpour et al., 1996, 1998).

3. Interferences

3.1. Contamination of soil samples may occur in either the sampling equipment or soil grinding equipment, especially for Zn and may occur with other metals also.

3.2. All apparatus that will come in direct contact with the extraction solution and extraction filtrate must be thoroughly washed and rinsed in pure hydrochloric acid (HCl) and pure water before use. Avoid contact with rubber and metal surfaces.

4. Precision and Accuracy

4.1. Repeated analysis of the same soil with medium concentration ranges of Cd, Cu, Ni, and Zn will give coefficients of variability from 10 to 15%. A major portion of the variance is related to the heterogeneity of the soil rather than to the extraction process or method of analysis.

5. Apparatus

5.1. No. 10 (2-mm opening) sieve.
5.2. 8.5-cm^3 scoop, volumetric.
5.3. Extraction flask, 125-mL polyethylene conical flasks.
5.4. Mechanical reciprocating shaker, 180 oscillations per minute.
5.5. Filter funnel, 11 cm, and funnel rack.
5.6. Whatman No. 42 ashless filter paper (or equivalent), 12.5 cm.
5.7. Atomic absorption spectrophotometer (AAS) or inductively coupled plasma emission spectrometer (ICP-AES).
5.8. pH meter with reproducibility to at least 0.1 pH unit and a glass electrode paired with a calomel reference electrode.
5.9. Analytical balance.
5.10. Volumetric flasks, pipettes and microburet as required for preparation of reagents and standard solutions.

6. Reagents (use reagent grade chemicals and pure water)

6.1. *Extraction Reagent* [DTPA (diethylenetriaminepentaacetic acid)]: Weigh 1.96 DTPA {[(HOCOCH$_2$)$_2$NCH$_2$]$_2$NCH$_2$COOH} into a 1,000-mL volumetric flask. Add 14.92 g triethanolamine (TEA). Bring to approximately 950 mL with water. Add 1.47 g calcium chloride (CaCl$_2$·2H$_2$O). Bring to 1,000-mL with water while adjusting the pH to exactly 7.3 with 6N hydrochloric acid (HCl). The final concentration will be 0.005 M DTPA (acid form), 0.1 M TEA, and 0.01 M CaCl$_2$.

6.2. *Cadmium Standard* (1,000 mg Cd L^{-1}): Commercially prepared standard, or weigh 1,000 g pure Cd metal into a 1,000-mL volumetric flask and add 5–10 mL conc. HCl. Evaporate almost to dryness and then dilute to 1,000 mL with Extraction Reagent (see 6.1). Prepare working standards by diluting aliquots of the 1,000 mg Cd L^{-1} standard with the Extraction Reagent (see 6.1) to cover the anticipated range in concentration to be found in the soil extraction filtrate. Working standards from 0.1 to 10 mg Cd L^{-1} should be sufficient for most soils.

6.3. *Copper Standard* (1,000 mg Cu L^{-1}): Commercially prepared standard, or weigh 1.000 g pure Cu metal into a 1,000-mL volumetric flask and add in minimum amount conc. nitric acid (HNO$_3$) and add 5 mL conc. HCl. Evaporate almost to dryness and dilute to 1,000 mL with Extraction Reagent (see 6.1). Prepare working standards by diluting aliquots of the 1,000 mg Cu L^{-1} standard with Extraction Reagent (see 6.1) to cover the anticipated range in concentration filtrate. An initial range of 0 to 100 mg Cu L^{-1} is suggested for sludge-amended soils.

6.4. *Nickel Standard* (1,000 mg Ni L^{-1}): Commercially prepared standard, or weigh 1.000 g pure Ni metal into a 1,000-mL volumetric flask and add in a minimum amount conc. HNO$_3$ and add 5 mL conc. HCl. Evaporate almost to dryness and dilute to 1,000 mL with Extraction Reagent (see 6.1). Prepare working standards by diluting aliquots of the 1,000 mg Ni L^{-1} standard with Extraction Reagent (see 6.1) to cover the anticipated range in concentration filtrate. Working standards from 0.1 to 10 mg Ni L^{-1} should be sufficient for most soils.

6.5. *Zinc Standard* (1,000 mg Zn L^{-1}): Commercially prepared standard, or weigh 1.000 g pure Zn metal into a 1,000-mL volumetric flask and add 5–10 mL conc. HCl. Evaporate almost to dryness and dilute to 1,000 mL with Extraction Reagent (see 6.1). Prepare working standards by diluting aliquots of the 1,000 mg Zn L^{-1} standard with Extraction Reagent (see 6.1) to cover the anticipated range in concentration to be found in the soil extraction filtrate. An initial range of 0 to 100 mg Zn L^{-1} is suggested for sludge-amended soils.

7. Procedure

7.1. *Extraction:* Weigh 10 g or measure 8.5-cm^3 (see 5.2) of air-dry <10-mesh (2-mm) soil into a 125-mL extraction flask (see 5.3), add 20 mL Extraction Reagent (see 6.1), and shake on a reciprocating shaker for two hours. Samples that are shaken longer than two hours will give high results because a final equilibrium of the metal and soil is not reached in two hours. Filter and collect the filtrate.

7.2. *Analysis:* The elements in the filtrate can be determined by either AAS or ICP-AES (see 2.3). Since instruments vary in their operating conditions, no specific details are given.

8. Calibration and Standards

8.1. *Working Standards:* Working standards should be prepared as described in Section 6. If element concentrations are found outside the range of the instrument or standards, suitable dilutions should be prepared starting with a 1:2 soil extract to Extraction Reagent (see 6.1) dilutions.

8.2. *Calibration:* Calibration procedures vary with instrument techniques and type of instrument. Every precaution should be taken to ensure that the proper procedures are taken and the manufacturer's recommendations are followed in the operation and calibration of the instrument used.

9. Calculation

9.1. The results are reported as kg ha^{-1} for a 20-cm depth. kg element ha^{-1} = mg element L^{-1} in extraction filtrate x 4. If the extraction filtrate is diluted, the dilution factor must be applied. To express the results in mg kg^{-1} of soil, use the following formula: mg kg^{-1} in soil = mg L^{-1} in solution × 2.

10. Effects of Storage

10.1. Soils may be stored in an air-dry condition for several months with no effects on the amount of Zn and Cu extracted, other elements not known. The influence of time and temperature is discussed by Houba and Novozamsky (1998).

11. Interpretation

11.1. Limited work has been done to evaluate soil test methods for sludge-amended soils. Rappaport et al. (1987, 1988), reported that the DTPA method correlated well with metals applied in sludge but found generally poor correlations with plant uptake. More research is needed in this area, particularly with sensitive crops.

12. Comments

12.1. Grinding can change the amount of DTPA-extractable heavy metals, especially Fe. Therefore, it is imperative that both grinding procedures and extraction procedures be standardized. The grinding process should approximate using a wooden roller to crush the soil aggregates (Soltanpour et al., 1976, 1979). The distribution of Fe, Zn, Mn, and Cu varies with particle size (Kahn, 1979), and wetting and drying will affect Fe, Zn, Mn, and Cu extraction (Kahn and Soltanpour, 1978).

9.3 REFERENCES

Adriano, D. C. 1986. *Trace Elements in the Terrestrial Environment.* New York: Springer-Verlag.

Amacher, M. C. 1990. Nickel, cadmium, and lead, In: *Methods of Soils Analysis, Part 3, Chemical Methods,* SSSA Book Series No. 5. D. L. Sparks (ed.) Madison, WI: Soil Science Society of America, 739–768.

Anon. 1998. *Heavy Metals in Soils and Phosphatic Fertilizers,* PPI/PPIC/FAR Technical Bulletin 1998-2. Norcross, GA: Potash and Phosphate Institute.

Baker, D. E. and M. C. Amacher. 1982. Nickel, copper, zinc, and cadmium, in: *Methods of Soil Analysis, Part 2,* Agronomy No. 9, A. L. Page (ed.). Madison, WI: American Society of Agronomy, 323–336.

Bartlett, R. J. and B. R. James. 1996. Chromium, In: *Methods of Soils Analysis, Part 3, Chemical Methods,* SSSA Book Series No 5. D. L. Sparks (Ed.). Madison, WI: Soil Science Society of America, 683–701.

Crock, J. G. 1996. Mercury, pp. 769–791. In: *Methods of Soils Analysis, Part 3, Chemical Methods,* SSSA Book Series No. 5, D.L. Sparks (ed.). Madison, WI: Soil Science Society of America, 769–791.

Holmgren, G. G. S., M. W. Meyer, R. L. Chaney, and R. B. Daniels. 1993. Cadmium, lead, zinc, copper, and nickel in agricultural soils in the United States of America. *J. Environ. Qual.* 22:335–348.

Houba, V. J. G. and I. Novozamsky. 1998. Influence of storage time and temperature of air-dried soils on pH and extractable nutrients using 0.01 M $CaCl_2$. *Fresenius J. Anal. Chem.* 360:362–365.

Huang, P. M. and R. Fujii. 1990. Selenium and arsenic, In: *Methods of Soils Analysis, Part 3, Chemical Methods,* SSSA Book Series No. 5. D. L. Sparks (ed.). Madison, WI: Soil Science Society of America, 793–831.

Johnson, G. V. and S. J. Donohue. 1992. Testing method for waste-amended soils, In: *Reference Soil and Media Diagnostic Procedures for the Southern Region of the United States,* Southern Cooperative Series Bulletin No. 374, S. J. Donohue (ed.). Blacksburg, VA: Virginia Agricultural Experiment Station, 46.

Kabata-Pendias, A. and D. C. Adriano. 1995. Trace metals, In: *Soil Amendments and Environmental Quality,* J. E. Rechcigl (ed.). Boca Raton, FL: Lewis Publishers, 139–167.

Kabata-Pendias, A. and H. Pendias. 1994. *Trace Elements in Soils and Plants,* 2nd ed. Boca Raton, FL: CRC Press.

Kahn, A. 1979. Distribution of DTPA-extractable Fe, Zn, and Cu in soil particle-size fractions. *Commun. Soil Sci. Plant Anal.* 10:1211–1218.

Kahn, A. and P. N. Soltanpour. 1978. Effect of wetting and drying on DTPA-extractable Fe, Zn, Mn, and Cu in soils. *Commun. Soil Sci. Plant Anal.* 9:193–202.

Lindsay, W. L. 1979. *Chemical Equilibria in Soils.* New York: John Wiley & Sons.

Lindsay, W. L. and W. A. Norvell. 1978. Development of a DTPA soil test for zinc, iron, manganese, and copper. *Soil Sci. Soc. Amer. Proc.* 42:421–428.

Pais, I. and J. B. Jones, Jr. 1997. *The Handbook of Trace Elements.* Boca Raton, FL: St. Lucie Press.

Radojevic, M. and V. N. Bashkin. 1999. *Practical Environmental Analysis.* Cambridge U.K.: The Royal Society of Chemistry.

Rappaport, B. D., J. D. Scott, D. C. Martens, R. B. Reneau, Jr., and T. W. Simpson. 1987. *Availability and Distribution of Heavy Metals, Nitrogen, and Phosphorus from Sewage Sludge in the Plant-Soil-Water Continuum.* VPI-VWRRC-Bull. 154 5C. Blacksburg, VA: Virginia Water Resources Research Center.

Rappaport, B. D., D. C. Martens, R. B. Reneau, Jr., and T. W. Simpson. 1988. Metal availability in sludge-amended soils with elevated metal levels. *J. Environ. Qual.* 17:42–47.

Risser, J. A. and D. E. Baker. 1990. Testing soils for toxic metals, In: *Soil Testing and Plant Analysis,* 3rd edition, SSSA Book Series No. 3, R. L. Westerman (ed.). Madison, WI: Soil Science Society of America, 275–298.

Soltanpour, P. N. 1991. Determination of nutrient availability and elemental toxicity by AB-DTPA soil test and ICPS. *Adv. Soil Sci.* 16:165–190.

Soltanpour, P. N., A. Khan, and W.L. Lindsay. 1976. Factors affecting DTPA-extractable Zn, Fe, Mn and Cu. *Commun. Soil Sci. Plant Anal.* 7:797–821.

Soltanpour, P. N. and A. P. Schwab. 1977. A new soil test for simultaneous extraction of macro- and micro-nutrients in alkaline soils. *Commun. Soil Sci. Plant Anal.* 8:195–207.

Soltanpour, P. N. and S. M. Workman. 1979. Modification of NH_4HCO_3-DTPA soil test to omit carbon black. *Commun. Soil Sci. Plant Anal.* 10:1411–1420.

Soltanpour, P. N., A. Khan, and A. P. Schwab. 1979. Effect of grinding variables on the NH_4HCO_3-DTPA soil test values for Fe, Zn, Mn, Cu, P, and K. *Commun. Soil Sci. Plant Anal.* 10:903–909.

Soltanpour, P. N., C. W. Johnson, S. M. Workman, J. B. Jones, Jr., and R. O. Miller. 1996. Inductively coupled plasma emission spectrometry and inductively coupled plasma-mass spectroscopy, In: *Methods of Soil Analysis, Part 3, Chemical Method.* SSSA Book Series No. 5. D. L. Sparks (ed.). Madison, WI: Soil Science Society of America, 91–139.

Soltanpour, P. N., C. W. Johnson, S. M. Workman, J. B. Jones, Jr., and R. O. Miller. 1998. Advances in ICP emission and ICP mass spectrometry. *Adv. Agron.* 64:28–113.

Soon, Y. K. and A. Abboud. 1993. Cadmium, chromium, lead, and nickel, In: *Soil Sampling and Methods of Analysis,* M.R. Carter (ed.). Boca Raton, FL: Lewis Publishers, 101–108.

Watson, M. E. and R. A. Isaac. 1990. Analytical instrumentation for soil and plant analysis, In: *Soil Testing and Plant Analysis,* 3rd ed., SSSA Book Series No. 3. R.L. Westerman (ed.). Madison, WI: Soil Science Society of America, 691–740.

Wright, R. J. and T. I. Stuczynski. 1996. Atomic absorption and flame emission spectrometry, In: *Methods of Soil Analysis, Part 3, Chemical Method,* SSSA Book Series No. 5. D. L. Sparks (ed.), Madison, WI: Soil Science Society of America, 65–90.

10 Ammonium- and Nitrate-Nitrogen

The presence and accumulation of nitrogen (N), primarily as nitrate-nitrogen (NO_3-N), is becoming of major interest from both crop production factors, yield, and product quality as well as environmental concerns from run-off and leaching into aquifers.

Commonly used test procedures for determining the NO_3- and ammonium (NH_4)-N levels in soils have been described by Houba et al. (1987), Dahnke and Johnson (1990), Johnson (1992), Maynard and Kalra (1993), Anon. (1994a), Griffin (1995), Mulvaney (1996), Gelderman and Beegle (1998), Radojevic and Bashkin (1999). Strong and Mason (1999) discuss the N characteristics of Australian soils and methods of analysis and interpretation.

Soil test interpretation methods have been described by Peck et al. (1977) and Dahnke and Johnson (1990), and the basis and summarization of N fertilizer recommendations are given by Black (1993), Muchovej and Rechcigl (1995), Ludwick (1997), Maynard and Hochmuth (1997), and Reid (1998). The Pre-Plant Soil Nitrate Test (PPNT) and Pre-Sidedress Nitrate Test (PSNT) are discussed by Gelderman and Beegle (1998), test procedures that are being used extensively to regulate the application of N fertilizer, primarily for corn.

10.1 AMMONIUM BICARBONATE–DTPA EXTRACTION (NITRATE DETERMINATION)

1. Principle of the Method

1.1. The Extraction Reagent is 1 M ammonium bicarbonate (NH_4HCO_3) in 0.005 M DTPA adjusted to a pH of 7.6 (Soltanpour and Schwab, 1977; Soltanpour and Workman, 1979; Soltanpour, 1991). The nitrate (NO_3) is soluble in any water-based solution.

2. Range, Sensitivity, and Analysis Procedure

2.1. The range and sensitivity are the same as those for other methods of NO_3 determination.
2.2. The NO_3 concentration in the extractant is determined by spectrophotometry (Watson and Isaac, 1990).

3. Interferences

3.1. The Extraction Reagent is unstable with regard to pH and must be kept under mineral oil to prevent a pH change. No interferences are likely if chloride (Cl) ions present are masked.

4. Precision and Accuracy

4.1. A coefficient of variability ranging from 5 to 10% can be expected for different determinations.

5. Apparatus

5.1. No. 10 (2-mm opening) sieve constructed from stainless steel or Nalgene.
5.2. Analytical balance.
5.3. 125-mL polyethylene conical flasks.
5.4. Eberbach reciprocating shaker (or equivalent), 180 oscillations per minute.
5.5. Whatman No. 42 filter paper (or equivalent), 12.5 cm.
5.6. UV-VIS spectrophotometer set at 420 nm.
5.7. 2.5-cm matching spectrophotometric tubes.
5.8. A pH meter readable to 0.01 pH units.
5.9. Funnels and funnel rack.
5.10. Accurate automatic diluter.
5.11. Volumetric flasks and pipettes as required for preparation of reagents and standard solutions.

6. Reagents (use reagent grade chemicals and pure water)

6.1. *Extraction Reagent* [Ammonium bicarbonate (AB)-DTPA]: 0.005 M DTPA (diethylenetriaminepentaacetic acid) solution is obtained by adding 9.85 g DTPA (acid form) to 4,500 mL water in a 5,000-mL volumetric flask. Shake for five hours constantly to dissolve the DTPA. Bring to 5,000 mL with water. This solution is stable with regard to pH. To 900 mL of the 0.005 M DTPA solution, add 79.06 g ammonium bicarbonate (NH_4HCO_3) gradually and stir gently with a rod to facilitate dissolution and to prevent effervescence when bicarbonate is added. The solution is diluted to 1,000 mL with the 0.005 M DTPA solution and mixed gently with a rod. The pH is adjusted to 7.6 with slow agitation with a rod by adding 2 M hydrochloric acid (HCl). The AB-DTPA solution must be stored under mineral oil. Check the pH after storage and adjust it with a 2 M HCl dropwise, if necessary. The cumulative volume of HCl added should not exceed 1 mL L^{-1} limit, after which a fresh solution should be prepared. Ideally, the extraction reagent is best made as needed.

6.2. *Antimony Sulfate Solution:* Weigh 0.5 g antimony metal (Sb) into 80 mL conc. sulfuric acid (H_2SO_4) and then add 20 mL water to the acid carefully to prevent splattering. Heating will facilitate the dissolution of antimony (Sb) metal in H_2SO_4. This solution is used for masking (complexing) chloride (Cl) in the NO_3 determination.

6.3. *Chromotropic Acid Solution* (CTA) [0.00137 M solution of CTA or 4,5-dihydroxy-2,7-naphthalene-disulfonic acid, disodium salt [$(HO)_2C_{10}H_4(SO_3Na)_2$]: weigh 0.5 g CTA in 4.0 kg conc. H_2SO_4. This solution is used to develop color with NO_3.

6.4. *Fisher G Carbon Black:* Used to eliminate organic matter interference.

7. Procedures

7.1. *Extraction Method:* Weigh 10 g air-dry soil <10-mesh (2-mm) into a 125-mL conical flask, add 20 mL Extraction Reagent (see 6.1), and shake on an Eberbach reciprocal shaker or an equivalent shaker for exactly 15 minutes at 180 cycles per minute with flasks kept open. The extracts are then filtered through Whatman 42 filter paper.

7.2. *Determination:* Mix 5 mL soil extract (see 7.1) with one 1-mL scoop of Fisher G Carbon Black (see 6.4), shake for five minutes or longer if required to decolorize the solution, and filter. Place a 0.5 mL aliquot of the decolorized soil extract, Working Standards (see 8.2) and AB-DTPA [0.00 NO_3-N solution (see 6.1)] into 2.5-cm matching spectrophotometric tubes. Add 3.0 mL water to each tube using an automatic diluter. Add 2.0 mL Antimony Sulfate Solution (see 6.2) followed by 6.5 mL CTA (see 6.3) in quick succession to each tube. Mix thoroughly and cool solution in tap water for consistent results. The spectrophotometer is set at zero absorbance at 420 nm using the 0.00 NO_3-N solution. The color intensity (absorbance) of soil extracts is read after two hours.

8. Standards

8.1. *Primary Nitrate-N Standard* (1,000 NO_3-N L^{-1}): Commercially prepared standard, or weigh 3.611 g potassium nitrate (KNO_3) into a 500-mL volumetric flask and bring to volume with water.

8.2. *Working Standards:* Pipette 0.25, 0.5, 1, 1.5, 2.5 mL aliquots of the Primary Nitrate-N Standard (see 8.1) into 100-mL volumetric flasks and bring to volume with Extraction Reagent (see 6.1) in order to obtain a series of standards containing 2.5, 5, 10, 15, and 25 mg NO_3-N L^{-1}, respectively.

9. Calculations

9.1. mg NO_3-N kg^{-1} in soil = mg NO_3-N L^{-1} in extract $\times$ 48.

10. Effects of Storage

10.1. Air-drying and storage should not have any significant effect on the level of NO_3 in the soil (Houba and Novozamsky, 1998).

10.2. The Extraction Reagent (see 6.1) must be stored under mineral oil.

11. Interpretation

11.1. Interpretation of the test results will depend on their use as described by Magdoff et al. (1984), Magdoff et al. (1990), Dahnke and Johnson (1990), Schmitt and Randall (1994), and Muchovej and Rechcigl (1995).

10.2 2 M POTASSIUM CHLORIDE EXTRACTION (NITRATE AND AMMONIUM DETERMINATION)

1. Principle of the Method

1.1. The Extraction Reagent is 2 M potassium chloride (KCl) (Dahnke and Johnson, 1990; Anon., 1994a; Mulvaney, 1996). The nitrate (NO_3) is soluble in any water-based solution and ammonium (NH_4^+) ions on colloidal exchange sites are brought into solution by exchange with the potassium (K^+) ion.

2. Range, Sensitivity, and Analysis Procedure

2.1. The range and sensitivity are the same as those for other methods of NO_3 and NH_4 determination.

2.2. The NO_3 concentration in the extractant is determined by a specific ion electrode (Mills, 1980; Johnson, 1992) and both NO_3 and NH_4 by spectrophotometric methods (Watson and Isaac, 1990; Mulvaney, 1996), or Kjeldahl distillation (Anon., 1994b).

3. Interferences

3.1. The Extracting Reagent is stable with regard to pH and no interferences are likely if chloride (Cl) ions present are masked.

4. Precision and Accuracy

4.1. A coefficient of variability ranging from 5 to 10% can be expected for different determinations.

5. Apparatus

5.1. No. 10 (2-mm opening) sieve.
5.2. Analytical balance.
5.3. 125-mL polyethylene conical flasks.
5.4. Eberbach reciprocating shaker (or equivalent), 180 oscillations per minute.
5.5. Whatman No. 2 filter paper (or equivalent), 11 cm.
5.6. UV-VIS spectrophotometer set at 420 nm.
5.7. 2.5-cm matching spectrophotometric tubes.
5.8. Nitrate specific-ion electrode and meter.
5.9. pH meter readable to 0.01 pH units.
5.10. Funnels and funnel rack.
5.11. Accurate automatic diluter.
5.12. Volumetric flasks and pipettes as required for preparation of reagents and standard solutions.

6. Reagents (use reagent grade chemical and pure water)

6.1. *Extraction Reagent* (2 M Potassium Chloride): Weigh 150 g potassium chloride (KCl) into a 1,000-mL volumetric flask and bring to volume with water.

Ammonium- and Nitrate-Nitrogen

6.2. *Antimony Sulfate Solution:* Weigh 0.5 g antimony metal (Sb) in 80 mL conc. sulfuric acid (H_2SO_4), and then add 20 mL water to the acid carefully to prevent splattering. Heating will facilitate the dissolution of antimony (Sb) metal in H_2SO_4. This solution is used for masking (complexing) chloride (Cl) in the NO_3 determination.

6.3. *Chromotropic Acid Solution* (CTA) [0.00137 M solution of CTA or 4,5-dihydroxy-2,7-naphthalene-disulfonic acid, disodium salt [$(HO)_2C_{10}H_4(SO_3Na)_2$]: Weigh 0.5 g CTA in 4.0 kg conc. H_2SO_4. This solution is used to develop color with NO_3.

6.4. *Fisher G Carbon Black:* used to eliminate organic matter interference.

7. Procedures

7.1. *Extraction Method:* Weigh 10 g air-dry soil <10-mesh (2-mm) into a 125-mL conical flask, add 50 mL Extraction Reagent (see 6.1), and shake on an Eberbach reciprocal shaker or an equivalent shaker for exactly 15 minutes at 180 cycles per minute. The extract is then filtered through Whatman 42 filter paper.

7.2. *Nitrate Determination:* Mix 5 mL of the soil extract (see 7.1) with one 1-mL scoop of Fisher G Carbon Black (see 6.4), shake for 5 minutes or longer if required to decolorize the solution, and filter. Place a 0.5 mL aliquot of the latter decolorized soil extract, Working Standards (see 8.2) and 2 M KCl [0.00 NO_3-N solution (see 6.1)] into 2.5-cm matching spectrophotometric tubes. Add 3.0 mL water to each tube using an automatic diluter. Add 2.0 mL Antimony Sulfate Solution (see 6.2), followed by 6.5 mL CTA (see 6.3) in quick succession to each tube. Mix thoroughly for consistent results. After two hours of cooling in tap water, the spectrophotometer is set at zero absorbance at 420 nm with the 0.00 NO_3-N solution. The color intensity (absorbance) of soil extracts is read after two hours.

7.3. *Ammonium Determination:* Follow the procedures as described by Dahnke and Johnson (1990), Mulvaney (1996), or the steam distillation procedure described in the WREP Bulletin 125 (Anon., 1994b)

8. Standards

8.1. *Primary Nitrate-N Standard* (1,000 NO_3-N L^{-1}): Commercially prepared standard, or weigh 7.22 g potassium nitrate (KNO_3) into a 1,000-mL volumetric flask and bring to volume with water.

8.2. *Working Nitrate-N Standards:* Pipette 0.25, 0.5, 1, 1.5, 2.5 mL aliquots of the Primary Nitrate-N Standard (see 8.1) into 100-mL volumetric flasks and bring to volume with Extraction Reagent (see 6.1) in order to obtain a series of standards containing 2.5, 5, 10, 15, and 25 mg NO_3-N L^{-1}, respectively.

8.3. *Primary Ammonium-N Standard* (1,000 mg NH_4-N L^{-1}): Commercially prepared standard, or weigh 4.72 g ammonium sulfate [$(NH_4)_2SO_4$] into a 1,000-mL volumetric flask and bring to volume with water.

8.4. *Working Ammonium-N Standards:* Pipette 0.25, 0.5, 1, 1.5, 2.5 mL aliquots of the Primary Ammonium-N Standard (see 8.3) into 100-mL volumetric flasks and bring to volume with Extraction Reagent (see 6.1) in order to obtain a series of standards containing 2.5, 5, 10, 15, and 25 mg NH_4-N L^{-1}, respectively.

9. Calculations

9.1. mg NO_3- or NH_4-N kg^{-1} in soil = mg NO_3 or NH_4-N L^{-1} in extract $\times$ 48.

10. Effects of Storage

10.1. Air-drying and storage should not have any significant effect on the level of NO_3 or NH_4 in the soil (Houba and Novozamsky, 1998).

11. Interpretation

11.1. Interpretation of the test results will depend on their use as described by Magdoff et al. (1984), Magdoff et al. (1990), Dahnke and Johnson (1990), Soltanpour (1991), Schmitt and Randall (1994) and Muchovej and Rechcigl (1995).

10.3 0.01 M CALCIUM SULFATE OR 0.04 M AMMONIUM SULFATE EXTRACTION (NITRATE DETERMINATION)

1. Principle of the Method

1.1. The Extraction Reagent is either 0.01 M calcium sulfate ($CaSO_4$) or 0.04 M ammonium sulfate [$(NH_4)_2SO_4$] as nitrate (NO_3) is soluble in any water-based solution.

2. Range, Sensitivity, and Analysis Procedure

2.1. The range and sensitivity are the same as those for other methods of NO_3 determination.
2.2. The NO_3 concentration in the extractant is determined by a specific ion electrode (Mills, 1980; Johnson, 1992) or spectrophotometric methods (Watson and Isaac, 1990; Mulvaney, 1996).

3. Interferences

3.1. The Extraction Reagents are stable with regard to pH and no interferences are likely if chloride (Cl) ions present are masked.

4. Precision and Accuracy

4.1. A coefficient of variability ranging from 5 to 10% can be expected for different determinations.

5. Apparatus

5.1. No. 10 (2-mm opening) sieve.
5.2. Analytical balance.
5.3. 125-mL polyethylene conical flasks.
5.4. Eberbach reciprocating shaker (or equivalent), 180 oscillations per minute.

5.5. Whatman No. 2 filter paper (or equivalent), 11 cm.
5.6. UV-VIS spectrophotometer set at 420 nm.
5.7. 2.5-cm matching spectrophotometric tubes.
5.8. Nitrate specific-ion electrode and meter.
5.9. pH meter readable to 0.01 pH units.
5.10. Funnels and funnel rack.
5.11. Accurate automatic diluter.
5.12. Volumetric flasks and pipettes as required for preparation of reagents, and standard solutions.

6. Reagents (use reagent grade chemical and pure water)

6.1. *Extraction Reagent* (0.01 M Calcium Sulfate): Weigh 1.72 g calcium sulfate dihydrate ($CaSO_4 \cdot 2H_2O$) into a 1,000-mL volumetric flask and bring to volume with water.
6.2. *Extraction Reagent* (0.04 M Ammonium Sulfate): Weigh 5.28 g ammonium sulfate [$(NH_4)_2SO_4$] into a 1,000-mL volumetric flask and bring to volume with water.
6.3. *Antimony Sulfate Solution:* Weigh 0.5 g antimony metal (Sb) in 80 mL conc. sulfuric acid (H_2SO_4), and then add 20 mL water to the acid carefully to prevent splattering. Heating will facilitate the dissolution of antimony (Sb) metal in H_2SO_4. This solution is used for masking (complexing) chloride (Cl) in the NO_3 determination.
6.4. *Chromotropic Acid Solution* (CTA) [0.00137 M solution of CTA or 4,5-dihydroxy-2,7-naphthalene-disulfonic acid, disodium salt [$(HO)_2C_{10}H_4(SO_3Na)_2$]: Weigh 0.5 g CTA in 4.0 kg conc. H_2SO_4. This solution is used to develop color with NO_3.
6.5. *Fisher G Carbon Black:* used to eliminate organic matter interference.

7. Procedures

7.1. *Extraction Method:* Weigh 5 g air-dry soil <10-mesh (2-mm) into a 125-mL conical flask, add 50 mL Extraction Reagent (see 6.1 or 6.2), and shake on an Eberbach reciprocal shaker or an equivalent shaker for exactly 15 minutes at 180 cycles per minute. The extract is then filtered through Whatman 2 filter paper.
7.2. *Determination:* Mix 5 mL soil extract (see 7.1) with one 1-mL scoop of Fisher G Carbon Black (see 6.5), shake for five minutes or longer if required to decolorize the solution, and filter. Place a 0.5 mL aliquot of the latter decolorized soil extract, Working Standards (see 8.2), and Extraction Reagent (see 6.1 or 6.2) into 2.5 cm matching spectrophotometric tubes. Add 3.0 mL water to each tube using an automatic diluter. Add 2.0 mL Antimony Sulfate Solution (see 6.3), followed by 6.5 mL CTA (see 6.4) in quick succession to each tube. Mix thoroughly for consistent results. After two hours of cooling in tap water, the spectrophotometer is set at zero absorbance at 420 nm with the 0.00 NO_3-N solution. The color intensity (absorbance) of soil extracts is read after two hours.

8. Standards

8.1. *Primary Nitrate-N Standard* (1,000 NO_3-N L^{-1}): Commercially prepared standard, or weigh 3.611 g potassium nitrate (KNO_3) into a 500-mL volumetric flask and bring to volume with water.

8.2. Working Standards: Pipette 0.25, 0.5, 1, 1.5, 2.5 mL aliquots of the Primary Nitrate-N Standard (see 8.1) into 100-mL volumetric flasks and bring to volume with Extraction Reagent (see 6.1 or 6.2) in order to obtain a series of standards containing 2.5, 5, 10, 15, and 25 mg NO_3-N L^{-1}, respectively.

9. Calculations

9.1. mg NO_3-N kg^{-1} in soil = mg NO_3-N L^{-1} in extract $\times$ 48.

10. Effects of Storage

10.1. Air-drying and storage should not have any significant effect on the level of NO_3 in the soil (Houba and Novozamsky, 1998).

11. Interpretation

11.1. Interpretation of the test results will depend on their use as described by Magdoff et al. (1984), Magdoff et al. (1990), Dahnke and Johnson (1990), Schmitt and Randall (1994), and Muchovej and Rechcigl (1995).

10.4 REFERENCES

Anon. 1994a. Ammonium- and nitrate-nitrogen: KCl extraction method, In: *Plant, Soils, and Water Reference Methods for the Western Region,* Western Regional Extension Publication WREP 125, R. G. Gavlak, D. A. Horneck, and R. O. Miller (eds.). Fairbanks, AK: University of Alaska, 48.

Anon. 1994b. Kjeldahl distillation: CEC, TN, NH_4-N, NO_3-N, and mineralizable-N, In: *Plant, Soils, and Water Reference Methods for the Western Region,* Western Regional Extension Publication WREP 125, R. G. Gavlak, D. A. Horneck, and R. O. Miller (eds.). Fairbanks, AK: University of Alaska, 45–47.

Black, C. A. 1993. *Soil Fertility Evaluation and Control.* Boca Raton, FL: Lewis Publishers.

Dahnke, W. C. and G. V. Johnson. 1990. Testing soils for available nitrogen, In: *Soil Testing and Plant Analysis,* 3rd ed., SSSA Book Series No. 3, R. L. Westerman (ed.). Madison, WI: Soil Science Society of America, 127–139.

Griffin, G., W. Jokela and D. Ross. 1995. Recommended soil nitrate-N tests, In: *Recommended Soil Testing Procedures for the Northeastern United States,* 2nd ed.. Northeast Regional Publication No. 493 (revised), J. T. Sims and A. M. Wolf (eds.). Newark, DE: University of Delaware.

Gelderman, R. H. and D. Beegle. 1998. Nitrate-nitrogen, In: *Recommended Chemical Soil Test Procedures for the North Central Region,* North Central Regional Research Publications No. 221 (revised), J. R. Brown (ed.). Columbia, MO: Missouri Agricultural Experiment Station SB 1001, University of Missouri, 17–20.

Houba, V. J. G., I. Novozamsky, J. Uittenbogaard, and J. J. van der Lee. 1987. Automatic determination of "total soluble nitrogen" in soil extracts. *Landw. Forschung* 40: 295–302.

Houba, V. J. G. and I. Novozamsky. 1998. Influence of storage time and temperature of air-dried soils on pH and extractable nutrients using 0.01 M $CaCl_2$. *Fresenicus J. Anal. Chem.* 360:362–365.

Johnson, G. V. 1992. Determination of nitrate-nitrogen by specific ion electrode, In: *Reference Soil and Media Diagnostic Procedures for the Southern Region of the United States,*

Southern Cooperative Series Bulletin No. 374, S. J. Donohue (ed.). Blacksburg, VA: Virginia Agricultural Experiment Station, 25–27.

Ludwick, A. E. (ed.). 1997. *Western Fertilizer Handbook, Second Horticulture Edition.* Danville, IL: Interstate Publishers, Inc.

Magdoff, F. R., D. Ross, and J. Amadon. 1984. A soil test for nitrogen availability to corn. *Soil Sci. Amer. J.* 48:1301–1305.

Magdoff, F. R., W. E. Jokela, R. H. Fox, and G. F. Griffin. 1990. A soil test for nitrogen availability in the northeastern U.S. *Commun. Soil Sci. Plant Anal.* 21:1102–1115.

Maynard, D. G. and Y. P. Kalra. 1993. Nitrate and exchangeable ammonium-nitrogen, In: *Soil Sampling and Methods of Analysis.* M. R. Carter (ed.). Boca Raton, FL: Lewis (CRC) Publishers, 25–38.

Maynard, D. N. and G. J. Hochmuth. 1997. *Knott's Handbook for Vegetable Growers*, 4th ed., New York: John Wiley & Sons.

Mills, H. A. 1980. Nitrogen specific-ion electrodes for soil, plant, and water analysis. *J. Assoc. Off. Anal. Chem.* 63:797–801.

Muchovej, R. M. C. and J. E. Rechcigl. 1995. Nitrogen fertilizers, In: *Soil Amendments and Environmental Quality,* J. E. Rechcigl (ed.). Boca Raton, FL: Lewis Publishers, 1–64.

Mulvaney, R. L. 1996. Nitrogen—Inorganic forms, In: *Methods of Soil Analysis, Part 3, Chemical Methods,* SSSA Book Series No. 5, R. L. Sparks (ed.). Madison, WI: Soil Science Society of America, 1123–1184.

Peck, T. R., J. T. Cope, Jr., and D. A. Whitney (eds.). 1977. *Soil Testing: Correlating and Interpreting the Analytical Results,* ASA Special Publication No. 29. Madison, WI: American Society of Agronomy.

Radojevic, M. and V. N. Bashkin. 1999. *Practical Environmental Analysis.* Cambridge, UK.: The Royal Society of Chemistry.

Reid, K. (ed.). 1998. *Soil Fertility Handbook.* Toronto: Ministry of Agriculture, Food and Rural Affairs. Queen's Printer for Ontario.

Schmitt, M. A. and G. W. Randall. 1994. Developing a soil test nitrogen test for improved recommendations for corn. *J. Prod. Agric.* 7:328–334.

Soltanpour, P. N. 1991. Determination of nutrient availability and elemental toxicity by AB-DTPA soil test and ICPS. *Adv. Soil Sci.* 16:165–190.

Soltanpour, P. N. and A. P. Schwab. 1977. A new soil test for simultaneous extraction of macro- and micro-nutrients in alkaline soils. *Commun. Soil Sci. Plant Anal.* 8:195–207.

Soltanpour, P. N. and S. M. Workman. 1979. Modification of NH_4HCO_3-DTPA soil test to omit carbon black. *Commun. Soil Sci. Plant Anal.* 10:1411–1420.

Strong, W. M. and M. G. Mason. 1999. Nitrogen, In: *Soil Analysis: An Interpretation Manual,* K. I. Peverill, L. A. Sparrow, and D. J. Reuter (eds.). Collingwood, Australia: CSIRO Publishing, 171–185.

Watson, M. E. and R. A. Isaac. 1990. Analytical instruments for soil and plant analysis, In: *Soil Testing and Plant Analysis,* 3rd ed. SSSA Book Series No. 3. R. L. Westerman (ed.). Madison, WI: Soil Science Society of America 691–740.

11 Sulfate-Sulfur

Sulfur (S) exists in soil in the form of primary minerals, such as sulfide minerals, pyrite and marcasite (FeS_2), chalcopyrite ($CuFeS_2$), and pyrrhotite ($Fe_{1-x}S$), as a component of soil organic matter, and as the sulfate (SO_4^{2-}) anion in the soil solution with a major portion of this form existing primarily in the subsoil. The SO_4^{2-} anion is easily absorbed by clay and iron and aluminum oxides, adsorption increasing with pH. The SO_4^{2-} anion is the primary form of S absorbed by plants, although S released from the primary minerals and soil organic matter will contribute to that available to plants. The depth of soil sampling will vary with method and those parameters used to interpret the analytical data (Lewis, 1999).

Three extraction procedures for determining SO_4-S are described (Johnson, 1992; Singh et al., 1995; Combs et al., 1998) plus the 0.01 M calcium chloride ($CaCl_2$) extraction method for total soluble S (Houba et al., 1990). Soil test interpretation methods are described by Peck et al. (1977) and Dahnke and Olson (1990), and the basis and summerization of S fertilizer recommendations are given by Johnson and Fixen (1990), Black (1993), Mikkelsen and Camberato (1995), Ludwick (1997), Maynard and Hochmuth (1997), and Reid (1998).

11.1 MONOCALCIUM PHOSPHATE EXTRACTION

1. Principle of the Method

1.1. Sulfate (SO_4) ions in the monocalcium phosphate extractant are precipitated by barium (Ba) ions and the produced turbidity measured spectrophotometrically at 420 nm. Other procedures for measuring the concentration of SO_4 in the extractant can be by ion chromatography (Anon., 1994) or inductively coupled plasma emission spectrometry (ICP-AES) (Novozamsky et al., 1986).

2. Range and Detection Limit

2.1. This procedure yields a standard curve that is linear up to 50 mg S L^{-1}.
2.2. The detection limit is approximately 0.05 mg S L^{-1} in the extract. The determination limit is approximately 0.15 mg S L^{-1} (1.5 mg S kg^{-1} in the soil).

3. Interferences

3.1. No interferences are expected.

4. Precision and Accuracy

4.1. The reproducibility of determinations by this procedure should give, with reasonable control and thorough sample preparation, a coefficient of variation within 8%.

5. Apparatus

5.1. No. 10 (2-mm opening) sieve.
5.2. Extraction vessel or flask, 50-mL.
5.3. Mechanical reciprocating shaker, minimum of 180 oscillations per minute.
5.4. Centrifuge with swing-out head.
5.5. Filter funnel, 11 cm.
5.6. Whatman No. 2 filter paper or equivalent, 11 cm.
5.7. Analytical balance.
5.8. Magnetic stirrer.
5.9. UV-VIS spectrophotometer set at 420 nm.
5.10. Spectrophotometric tubes or cuvettes.
5.11. Volumetric flasks and pipettes as required for preparation of reagents and standard solutions.

6. Reagents (use reagent grade chemicals and pure water)

6.1. *Extraction Reagent* [Calcium Phosphate solution (500 mg L^{-1})]: Weigh 2.03 g calcium phosphate [$Ca(H_2PO_4)_2 \cdot 2H_2O$] into a 1,000-mL volumetric flask and bring to volume with water.

6.2. *Seed Solution:* Weigh 0.1087 g potassium sulfate (K_2SO_4) in 500 mL water and add 500 mL conc. hydrochloric acid (HCl). Stir to bring into solution and slowly add 2.0 g powdered gum acacia with stirring to bring into solution. Keep the prepared solution refrigerated.

6.3. *EDTA Solution (0.02 M):* Weigh 5.84 g ethylenediaminetetraacetic acid (H_4EDTA) into a 1,000-mL volumetric flask, add 30 mL conc. aqueous ammonia (NH_4OH, s.p. = 0.91 g cm^{-3}), and bring to volume with water.

6.4. *Barium Chloride Crystals:* Crush barium chloride ($BaCl_2 \cdot H_2O$) to pass as 20- to 30-mesh sieve.

6.5. *Charcoal:* Wash Darco G-60 Activated Carbon with Extraction Reagent (see 6.1) until free of measureable S. Rinse with water and oven dry. Store in a closed container.

6.6. *Stock Solution* (1,000 mg S L^{-1}): Commercially prepared standard, or weigh 5.434 g potassium sulfate (K_2SO_4) into a 1,000-mL volumetric flask in about 400 mL EDTA Solution (see 6.3) and bring to volume with EDTA Solution (see 6.3).

7. Extraction Procedure

7.1. *Extraction:* Weigh 10 g air-dry <10-mesh (2-mm) soil into an extraction vessel (see 5.2), pipette into the vessel 25 mL Extraction Reagent (see 6.1), and shake for 30 minutes. Add 1/4-teaspoon (about 0.15 g) powdered Charcoal (see 6.5) and shake for an additional 3 minutes. Filter and transfer a 10 mL aliquot into another flask.

8. Analytical Procedure

8.1. *Measurement:* To a 10 mL aliquot (see 7.1), add 1 mL Seed Solution (see 6.2), and immediately swirl the contents. Place the flask on magnetic stirrer and add

Sulfate-Sulfur

1/4-teaspoon (0.3 g) Barium Chloride Crystals (see 6.4). Stir for about one minute. Transfer an aliquot to a spectrophotometric tube or cuvette (see 5.10) and then read the absorbance using a calibrated UV-VIS spectrophotometer (see 5.9) at 420 nm.

9. Calibration and Standards

9.1. *Standard Series:* Pipette 0, 1, 2, 3, 4, and 5 mL Stock Solution (see 6.6) into a series of 100-mL volumetric flasks and bring to volume with EDTA Solution (see 6.3). This Standard Series will have S concentrations of 0, 10, 20, 30, 40, and 50 mg L^{-1}, respectively.

9.2 *Calibration Curve:* Plot the emission counts versus mg S L^{-1} in the Standard Series.

10. Calculation

10.1. The SO_4-S content of the soil material, expressed in mg SO_4-S kg^{-1} is determined by multiplying the SO_4-S content in the extractant (see 7.1) by 2.5.

11. Interpretation

11.1. Suggested critical range is 5 to 10 mg extractable S kg^{-1} (Lewis, 1999).

12. Effects of Storage

12.1. Not yet known.

11.2   0.5 M AMMONIUM ACETATE–0.25 M ACETIC ACID EXTRACTION

1. Principle of the Method

1.1. Sulfate (SO_4) ions in the 0.5 M ammonium acetate–0.25 M acetic acid extractant are precipitated by barium (Ba) ions and the precipitate produced turbidity measured spectrophotometrically at 420 nm. Other procedures for measuring the concentration of SO_4 in the extractant are by ion chromatography (Anon., 1994) or inductively coupled plasma emission spectrometry (ICP-AES) (Novozamsky et al., 1986).

2. Range and Detection Limit

2.1. This procedure yields a standard curve that is linear up to 50 mg S L^{-1}.
2.2. The detection limit is approximately 0.05 mg S L^{-1} in the extract. The determination limit is approximately 0.15 mg S L^{-1} (1.5 mg S kg^{-1} in the soil).

3. Interferences

3.1. No interferences are expected.

4. Precision and Accuracy

4.1. The reproducibility of determinations by this procedure should give, at reasonable control and thorough sample preparation, a coefficient of variation within 8%.

5. Apparatus

5.1. No. 10 (2-mm opening) sieve.
5.2. Extraction vessel or flask, 50-mL.
5.3. Mechanical reciprocating shaker, minimum of 180 oscillations per minute.
5.4. Centrifuge with swing-out head.
5.5. Filter funnel, 11 cm.
5.6. Whatman No. 2 filter paper or equivalent, 11 cm.
5.7. Analytical balance.
5.8. Magnetic stirrer.
5.9. UV-VIS spectrophotometer set at 420 nm.
5.10. Spectrophotometric tubes or cuvettes.
5.11. Volumetric flasks and pipettes as required for preparation of reagents and standard solutions.

6. Reagents (use reagent grade chemicals and pure water)

6.1. *Extraction Reagent* (0.5 M ammonium acetate–0.25 M acetic acid): Weigh 39 g ammonium acetate ($NH_4C_2H_3O_2$) into a 1,000-mL volumetric flask and bring to volume with 0.25 M acetic acid ($HC_2H_3O_2$) (dilute 14.31 glacial $HC_2H_3O_2$ in 1,000 mL water).

6.2. *Seed Solution:* Dissolve 0.1087 g potassium sulfate (K_2SO_4) in 500 mL water and add 500 mL conc. hydrochloric acid (HCl). Stir to bring into solution and slowly add 2.0 g powdered gum acacia with stirring to bring into solution. Keep the prepared solution refrigerated.

6.3. EDTA Solution (0.02 M): Weigh 5.84 g ethylenediaminetetraacetic acid (H_4EDTA) into a 1,000-mL volumetric flask, add 30 mL conc. aqueous ammonia (NH_4OH, s.p. = 0.91 g cm^{-3}), and bring to volume with water.

6.4. *Barium Chloride Crystals:* Crush barium chloride ($BaCl_2 \cdot H_2O$) to pass as 20- to 30-mesh sieve.

6.5. *Charcoal:* Wash Darco G-60 Activated Carbon with Extraction Reagent (see 6.1) until free of measureable S. Rinse with water and oven dry. Store in a closed container.

6.6. *Stock Solution* (1,000 mg S L^{-1}): Commercially prepared standard, or weigh 5.434 g potassium sulfate (K_2SO_4) into a 1,000-mL volumetric flask in about 400 mL EDTA Solution (see 6.3) and bring to volume with EDTA Solution (see 6.3).

7. Extraction Procedure

7.1. *Extraction:* Weigh 10 g air-dry <10-mesh (2-mm) soil into an extraction vessel, pipette
25 mL Extraction Reagent (see 6.1) into the flask, and shake for 30 minutes. Add 1/4-teaspoon (about 0.15 g) of powdered Charcoal (see 6.5) and shake for an additional 3 minutes. Filter and transfer a 10-mL aliquot into another flask.

8. Analytical Procedure

8.1. *Measurement:* To a 10-mL aliquot (see 7.1), add 1 mL Seed Solution (see 6.2) and immediately swirl the contents. Place the flask on magnetic stirrer and add 1/4-teaspoon (0.3 g) Barium Chloride Crystals (see 6.4). Stir for about 1 minute. Transfer an aliquot to a spectrophotometric tube or cuvette (see 5.10) and then read the absorbance using a calibrated UV-VIS spectrophotometer (see 5.9) at 420 nm.

9. Calibration and Standards

9.1. *Standard Series:* Pipette 0, 1, 2, 3, 4, and 5 mL Stock Solution (see 6.6) into a series of 100-mL volumetric flasks and bring to volume with EDTA Solution (see 6.3). This Standard Series has S concentrations of 0, 10, 20, 30, 40, and 50 mg S L^{-1}, respectively.

9.2. *Calibration Curve:* Plot the emission counts versus mg S L^{-1} in the Standard Series.

10. Calculation

10.1. The SO_4-S content of the soil material, expressed in mg SO_4-S kg^{-1} is determined by multiplying the SO_4-S content in the extractant (see 7.1) by 2.5.

11. Effects of Storage

11.1. Not yet known.

11.3 REFERENCES

Anon. 1994. Sulfate-sulfur (SO_4-S): Ion chromatographic method, In: *Plant, Soils, and Water Reference Methods for the Western Region,* Western Regional Extension Publication WREP 125, R. G. Gavlak, D. A. Horneck, and R. O. Miller (eds.). Fairbanks, AK: University of Alaska, 49–50.

Black, C. A. 1993. *Soil Fertility Evaluation and Control.* Boca Raton, FL: Lewis Publishers.

Combs, S., J. Denning, and K. D. Frank. 1998. Sulfate-sulfur, In: *Recommended Chemical Soil Test Procedures for the North Central Region,* North Central Regional Research Publication 221 (revised) J. R. Brown (ed.). Columbia, MO: Missouri Agricultural Experiment Station SB, University of Missouri, 35–40.

Dahnke, W. C. and R. A. Olson. 1990. Soil test correlation, calibration, and recommendations, In: *Soil Testing and Plant Analysis,* 3rd ed., SSSA Book No. 3, R. L. Westerman (ed.). Madison, WI: Soil Science Society of America, 45–71.

Houba, V .J. G., I. Novozamsky, Th. M. Lexmond, and S. J. van der Lee. 1990. Applicability of 0.01 M $CaCl_2$ as a single extractant for the assessment of the nutrient status of soils and other diagnostic purposes. *Commun. Soil Sci. Plant Anal.* 21:2281–2290.

Johnson, G. V. 1992. Determination of sulfate-sulfur by moncalcium phosphate extraction, In: *Reference Soil and Media Diagnostic Procedures for the Southern Region of the United States,* Southern Cooperative Series Bulletin No. 374, S. J. Donohue (ed.). Blacksburg, VA: Virginia Agricultural Experiment Station, 13–15.

Johnson, G. V., and P. E. Fixen. 1990. Testing soils for sulfur, boron, molybdenum, and chlorine, In: *Soil Testing and Plant Analysis, Third Ed.,* SSSA Book Series No. 3, R. L. Westerman (ed.). Madison, WI: Soil Science Society of America, 265–273.

Lewis, D. C. 1999. Sulfur, In: *Soil Analysis: An Interpretation Manual.* K. I. Peverill, L. A. Sparrow, and D. J. Reuter (eds.). Collingwood, Australia: CSIRO Publishing, 221–228.

Ludwick, A. E. (ed.). 1997. *Western Fertilizer Handbook: Second Horticulture Edition.* Danville, IL: Interstate Publishers, Inc.,

Maynard, D. N. and G. J. Hochmuth. 1997. *Knott's Handbook for Vegetable Growers,* 4th ed. New York; John Wiley & Sons,

Mikkelsen, R. L. and J. J. Camberato. 1995. Potassium, sulfur, lime, and micronutrient fertilizers, In: *Soil Amendments and Environmental Quality,* J. E. Rechcigl (ed.). Boca Raton, FL: Lewis Publishers, 109–137.

Novozamsky, I., R. van Eck, J. J. van der Lee, V. J. G. Houba, and E. J. M. Temminghoff. 1986. Determination of total sulphur and extractable sulfate in plant materials by inductively coupled plasma atomic emission spectrometry. *Commun. Soil Sci. Plant Anal.* 17:1147–1157.

Peck, T. R., J. T. Cope, Jr., and D. A. Whitney (eds.). 1977. *Soil Testing: Correlating and Interpreting the Analytical Results,* ASA Special Publication No. 29. Madison, WI: American Society of Agronomy.

Reid, K. (ed.). 1998. *Soil Fertility Handbook.* Toronto: Ontario Ministry of Agriculture, Food and Rural Affairs, Queen's Printer for Ontario.

Singh, Rabinder, D. K. Bhumbia, and R. F. Keefer. 1995. Recommended soil sulfate-S tests, In: *Recommended Soil Testing Procedures for the Northeastern United States,* Northeast Regional Publication No. 493 (revised), J. T. Sims and A. M. Wolf (eds.). Newark, DE: University of Delaware, 46–50.

12 Chloride

Although chlorine (Cl) has been identified as an essential micronutrient element since 1954 (Glass, 1989), there has been little focus on its soil determination for deficiency with more concern associated with its excess in saline soils or where salinity conditions exist from the use of saline irrigation water (Richards, 1954).

Historically, soil Cl analysis has been conducted primarily for the purpose of salinity characterization and irrigation management. However, recent research in the northwestern United States and in the northern Great Plains has indicated positive cereal responses to Cl additions (Christensen et al., 1981; Timm et al., 1986; Fixen et al., 1986a, 1986b, 1987). Studies in South Dakota have indicated that soil Cl level is a factor influencing the probability of obtaining a yield response to Cl (Fixen et al., 1987). The procedures that will be discussed here are intended for determining Cl fertilizer needs rather than for salinity evaluation (Johnson and Fixen, 1990; Gelderman et al., 1998). Therefore, detection of relatively low Cl concentrations is emphasized.

Chloride is similar to nitrate (NO_3) in solubility and mobility in the soil and, like NO_3, requires sampling depths greater than six inches. Fixen et al. (1987) found that a two-foot sampling depth was superior to shallower or deeper depths for predicting wheat plant Cl concentrations in eastern South Dakota.

Chloride is a ubiquitous ion and precautions must be taken to avoid contamination during sampling and in the laboratory. Many common laboratory reagents and cleansers contain Cl. Other possible sources of contamination include dust, perspiration, filter paper, glassware, water, and paper bags (Parker et al., 1983). Plastic gloves should be worn when handling filter paper for Cl determination.

Standard soil sample preparation procedures, as discussed in chapter one, appear to be adequate for Cl determination. Considerable flexibility exists in extraction techniques. Extractants that have been used include water (Fixen et al., 1987), 0.01 M calcium nitrate [$Ca(NO_3)_2$] (Bolton, 1971), 0.1 M sodium nitrate ($NaNO_3$) (Gaines et al., 1984), and 0.5M potassium sulfate (K_2SO_4). Theoretically, these should give similar results; however, the method of determination used may make some extractants more convenient than others. Time required for extraction appears to be similar to nitrate extraction. Gaines et al. (1984) showed that a 15-minute extraction on Georgia soils was adequate. Other investigators have adopted longer extraction periods of 15 (Fixen et al., 1987), 30, or 60 (Bolton, 1971) minutes. Minimum extraction times should be determined through recovery studies on the soils to be analyzed.

Several methods have been developed to determine Cl in soil extracts (Adriano and Doner, 1982; Sims and Johnson, 1991; Frankenberger et al., 1996). Many of these are not suited for routine soil testing and will not be discussed here. The procedures presented are those that have been successfully used in soil testing laboratories in the North Central Region (Gelderman et al., 1998).

Wheat will respond to Cl fertilization when the soil test level in the top 12 inches is less than 30 lbs Cl A^{-1}, and a probable response between 30 to 60 lbs Cl A^{-1}. There is no expected response when greater than 60 lbs Cl A^{-1}. The recommended rate is 40 lbs Cl A^{-1} as murite of potash (KCl).

12.1 0.01 M CALCIUM NITRATE EXTRACTION

1. Principle of the Method

1.1. This method is a modification of the procedure of Adriano and Doner (1982) for chloride (Cl) determination and uses an extraction procedure similar to that suggested by Bolton (1971). In this spectrophotometric method, Cl displaces thiocyanate which, in the presence of ferric (Fe^{3+}) iron, forms a highly colored ferric thiocyanate complex:

$$2\,Cl^- + Hg(SCN)_2 + 2\,Fe^{3+} \rightarrow HgCl_2 + 2\,Fe(SCN)^{2+}.$$

The color of the resulting solution is stable and proportional to the original Cl$^-$ ion concentration.

2. Range and Sensitivity

2.1. The procedure is very sensitive and has a detection limit of approximately 1 µg Cl g$^-$ soil.

3. Interferences

3.1. Nitrate (NO$_3^-$), sulfide (S$^-$), cyanide (CN$^-$), thiocyanate (CSN$^-$), bromide (Br$^-$), and iodide (I$^-$) anions can cause interferences but are usually not present in sufficient amounts to be a problem. Similar procedures have been modified for use with AutoAnalyzers.

4. Precision and Accuracy

4.1. Precision varies with level in the soil with coefficients from 9 to 24% with Cl levels greater than 12 mg kg^{-1} and 15 to 25% for Cl levels less than 10 mg kg^{-1} (Gelderman et al., 1998).

5. Apparatus

5.1. Standard NCR-13 10-g scoop.
5.2. UV-VIS spectrophotometer set at 460 nm.
5.3. Spectrophotometric tube or cuvet.
5.4. Shaker, capable of 180 or more excursions per minute (epm).
5.5. 50-mL Erlenmeyer flasks and filter funnels or tubes.

Chloride

6. Reagents (use reagent grade chemicals and pure water)

6.1. *Extraction Reagent* (0.01M Calcium Nitrate): Weigh 4.72 g calcium nitrate [Ca(NO$_3$)$_2$·4H$_2$O] into a 2,000-mL volumetric flask and bring to volume with water.

6.2. *Saturated Mercury (II) Thiocyanate Solution* (0.075%): Weigh approximately 0.75 g mercury thiocyanate [Hg(SCN)$_2$] into 1,000 mL of water and stir overnight. Filter through Whatman No. 42 paper. It is important that this solution be saturated because it can then be stored over a long period of time.

6.3. *Ferric (III) Nitrate Nonohydrate Solution:* Weigh 20.2 g ferric nitrate [Fe(NO$_3$)$_3$·9H$_2$O] in approximately 500 mL of water in a 1,000-mL volumetric flask and add conc. nitric acid (HNO$_3$) until the solution is almost colorless (20 to 30 mL). Bring to volume with water. Excess HNO$_3$ is not important so long as the amount is sufficient enough to prevent darkening of the solution when stored.

6.4. *Chloride Standard Stock Solution* (1,000 mg Cl L^{-1}): Commercially prepared standard, or weigh 0.2103 g potassium chloride (KCl) into a 100-mL volumetric flask and bring to volume with Extraction Reagent (see 6.1).

6.5. *Chloride Standard Intermediate Solution* (100 mg Cl L^{-1}): Pipette 10 mL Stock Solution (see 6.4) into a 100-mL volumetric flask and bring to volume with Extraction Reagent (see 6.1).

6.6. *Chloride Standard Working Solutions:* Pipette 0.5, 1.0, 2.0, 3.0, 4.0, 5.0, and 10.0 mL 100 mg Cl L^{-1} Standard Intermediate Solution (see 6.5) into 100-mL volumetric flasks and bring to volume with Extraction Reagent (see 6.1). These working standards contain 0.5, 1.0, 2.0, 4.0, 5.0, and 10.0 mg Cl L^{-1}, respectively.

6.7. *Charcoal:* Wash in 0.01 M Ca(NO$_3$)$_2$ and dry.

7. Procedure

7.1. *Extraction:* Scoop 10 g crushed soil (see 5.1) into a 50-mL Erlenmeyer flask, add approximately 25 mg 0.01 M Ca(NO$_3$)$_2$-washed charcoal (see 6.7), 25 mL Extraction Reagent (see 6.1), and shake for 15 minutes at 180 or more epm. Filter immediately following shaking using Whatman No. 42 filter paper or equivalent. Transfer 10 mL aliquot to a 50-mL beaker. Do duplicate or triplicate analyses and include a blank sample.

7.2. *Color Development:* To 10 mL aliquots of the extractant (see 7.1), Working Standards (see 6.6), and blank(s), add 4 mL each of Hg(SCN)$_2$ (see 6.2) and Fe(NO$_3$)$_3$·9H$_2$O (see 6.3). Swirl to mix. Allow 10 minutes for color development and then measure the color intensity using a VIS-UV spectrophotometer set at 460 nm. Set 100% T with Extracting Reagent (see 6.1).

8. Calibration

8.1. *Calibration curve:* Prepare a standard curve by pipetting 10 mL aliquots of each of the Working Standards (see 6.6) and proceed with color development (see 7.2). Plot transmittance or absorbance against concentration of the working standards.

9. Calculation

9.1. Determine Cl concentration in the extract from the meter reading and standard curve. Subtract the Cl in the blank and convert to mg Cl kg^{-1} in soil by multiplying by the dilution factor of 2.5.

10. Effects of Storage

10.1. Not known.

12.2 0.5 M POTASSIUM SULFATE EXTRACTION

1. Principle of the Method

1.1. Direct reading of soil extracts with the solid-state chloride (Cl) electrode has not been reliable across diverse soils and may give high readings (Hipp and Langdale, 1971). The electrode has worked well when used as an endpoint indicator in titrations. A more convenient alternative to potentiometric titrations is the potentiometric known addition method outlined here. It is particularly well suited for situations where occasional analysis for Cl concentration is needed since no calibration is necessary.

1.2. The basic approach of the method was reported by Bruton (1971) for Cl and fluoride (F) determination and involves measuring the electrode potential before and after addition of a known quantity of Cl to a sample. The change in potential is then related to sample concentration by assuming a Nernst-type relationship and a theoretical electrode response of 59.1 mV per tenfold change in concentration. This electrode response should be verified by measuring the potential after successive additions of the standard.

2. Range and Sensitivity

2.1. Determined by the characteristics of the electrode/meter system (Hipp and Langdale, 1971; Watson and Isaac, 1990).

3. Interferences

3.1. Not known.

4. Precision and Accuracy

4.1. Precision varies with level in the soil with coefficients from 9 to 24% with Cl levels greater than 12 mg kg^{-1}, and 15 to 25% for Cl levels less than 10 mg kg^{-1} (Gelderman et al., 1998).

5. Apparatus

5.1. Standard NCR-13 10-g scoop.
5.2. Shaker, capable of 180 or more excursions per minute (epm).

Chloride

5.3. Solid-state chloride electrode and double junction reference electrode.
5.4. pH/ion meter or pH-millivolt meter.
5.5. Magnetic stirrer.

6. Reagents (use reagent grade chemicals and pure water)

6.1. *Extraction Reagent* (0.5 M Potassium Sulfate): Weigh 87.0 g potassium sulfate (K_2SO_4) into a 1,000-mL volumetric flask and bring to volume with water.
6.2. *Chloride Standard Stock Solution* (1,000 mg Cl L^{-1}): Commercially prepared standard, or weigh 0.2103 g potassium chloride (KCl) into a 100-mL volumetric flask and bring to volume with Extraction Reagent (see 6.1).
6.3. *Chloride Standard Working Solution* (50 mg Cl^{-1}): Pipette 5 mL Stock Solution (see 6.2) into a 100-mL volumetric flask and bring to volume with Extraction Reagent (see 6.1).

7. Procedure

7.1. *Extraction:* Scoop 10 g crushed soil (see 5.1) nto 50-mL Erlenmeyer flask, add 30 mL Extraction Reagent (see 6.1), and shake for 15 minutes at 180 or more epm. Samples can be either filtered (Whatman No. 2 or equivalent), centrifuged, or left to settle to produce clear solutions. Do duplicate or triplicate analyses, and include a blank sample.

7.2. *Determination:* Pipette 20 mL of the extractant (see 7.1) into a 50-mL beaker. Place beaker on a stirrer, add magnet, and mix. Immerse chloride electrode into the beaker and record MV reading once the meter has stabilized. Add 2 mL 50 mg Cl L^{-1} solution and record MV reading when meter has stabilized. The difference between the first and second readings is ΔE. Sample concentration can be determined by either of the following approaches.

7.2.1. Obtain a Q value which corresponds to the ΔE value from a known addition table that is usually supplied with the electrode. Multiply the Q value by the concentration of the standard (mg 50 Cl L^{-1}) and subtract the blank concentration to determine the sample concentration, or

7.2.2. Calculate the concentration directly as follows:

$$C = (C_s)(V_s)/(V + V_s)\{10^{-\Delta E/59.1} - [V/(V + V_s)]\}^{-1}$$

where: C = concentration of sample
C_s = concentration of standard
V = mL of sample
V_s = mL of standard

In this procedure the equation simplifies to:

$$C = 4.545/10^{-\Delta E/59.1} - 0.909$$

Subtract the blank concentration from C.

8. Calculation

8.1. Convert extract concentration to mg kg^{-1} in soil by multiplying by the dilution factor of 3.0.

9. Effects of Storage

9.1. Not known.

12.3 SATURATED CALCIUM HYDROXIDE EXTRACTION

1. Principle of the Method

1.1. Chemically suppressed ion chromatography was introduced by Small et al., (1975). The main advantages of this method are high sensitivity, the ability to separate and quantify similar types of ions [i.e., flouride (F), chloride (Cl), and bromide (Br)], multiple element analyses and increased freedom from sample matrix effect. Mosko (1984) demonstrated some problems encountered in the analyses of a range of aqueous samples. The method of extraction of Cl from the soil is the same method used for nitrate (NO_3) (Carson, 1980). This allows the potential of multielement analyses.

2. Range and Sensitivity

2.1. The procedure is very sensitive and has a detection limit of approximately 1 μg Cl^{-1} soil.

3. Interferences

3.1. Minimal for this analytical technique.

4. Precision and Accuracy

4.1. Precision varies with level in the soil with coefficients from 9 to 24% with Cl levels greater than 12 mg kg^{-1} and 15 to 25% for Cl levels less than 10 mg kg^{-1} (Gelderman et al., 1998).

5. Apparatus

5.1. Balance (sensitivity 0.01 g).
5.2. Reciprocating shaker capable of approximately 200 epm.
5.3. Dispenser or buret capable of dispensing 25 mL.
5.4. 50-mL Erlenmeyer flasks and filter-funnel tubes.
5.5. Mechanical vacuum extractor (Centurion) and syringes.
5.6. Ion chromatography system including appropriate inline filters, column(s) and detector.

Chloride 171

5.7. Strip chart recorder and/or microcomputer aided data aquisition.

6. Reagents (use reagent grade chemicals and pure water)

6.1. *Extraction Reagent* (Saturated Calcium Hydroxide): Weigh 3 g calcium oxide (CaO) into 1,000-mL of water, and shake thoroughly. Filtration of the solution is desirable, but not necessary.

6.2. *Eluant for Ion Chromatograph:* Weigh 0.2544 g sodium carbonate (Na_2CO_3) and 0.2520 g sodium bicarbonate ($NaHCO_3$) into a 1,000-mL volumetric flask and bring to volume with water. Regenerant for chemically suppressed ion chromatography system utilizing a micromembrane suppressor. Add 1.5 mL conc. sulfuric acid (H_2SO_4) to a 1,000-mL volumetric flask and bring to volume with water.

6.3. *Chloride Stock Standard Solution* (1,000 mg Cl L^{-1}): Commercially prepared standard, or weigh 0.1648 g sodium chloride (NaCl) into a 100-mL volumetric flask and bring to volume with Extraction Reagent (see 6.1).

6.4. *Chloride Standard Intermediate Solution* (100 mg Cl L^{-1}): Pipette 10 mL Chloride Stock Solution (see 6.3) into a 100-mL volumetric flask and bring to volume with Extraction Reagent (see 6.1).

6.5. Chloride Working Standards (Ion Chromatographic Method)

Beaker	Volume of 100 mg Cl L^{-1} mL	Final volume mL	Chloride conc. in solution mg Cl L^{-1}	Equiv. conc. in soil mg Cl kg^{-1}
1	0.0	100	0.0	0.0
2	0.5	100	0.5	1.25
3	1.0	100	1.0	2.5
4	5.0	100	5.0	12.5
5	10.0	100	10.0	25.0
6	20.0	100	20.0	50.0

6.6. *Ion Exchange Resin:* Dowex© 50W-XB or equivalent (Biorad© 50W-XS) 50- to 100-mesh (prevents divalent and trivalent cations poisoning of separator and suppressor columns).

7. Procedure

7.1. *Extraction:* Weigh 10.0 g crushed soil into a 50-mL Erlenmeyer flask and weigh approximately 0.1 to 0.2 g calcium oxide (CaO) into each flask. Dispense 25 mL water into each flask and shake for five minutes at 180 or more epm. Filter sample into filter tubes through Whatman No. 2 filter paper that has been washed with water. Set up mechanized vacuum extractor utilizing 0.2-micron filters. Pour the filtered sample extract or standard into each syringe and allow it to equilibrate with the exchange resin about five minutes. Extract samples and/or standards through 0.2-micron filters.

7.2. *Ion Chromatographic Analysis Parameters:*

   a. Eluant flow rate.
   b. Regenerant flow rate.
   c. Retention time for Cl (determined experimentally).

7.3. *Analysis:* Inject samples and using the strip chart recorder and/or microcomputer to acquire the chromatographic data. Results are calculated by measuring peak height or peak area from strip chart recording or by computer software when sample run is complete.

   a. The fit of the standards to the standard calibration curve must be checked.
   b. Dilution factors are checked and data checked for error.
   c. Quality control standards (soil extracts and solution standards) are checked.

7.4. *Cleaning Laboratory Ware:*

   a. Rinse with water.
   b. Soak in a dilute acid bath (at least 30 minutes).
   c. Rinse three times with water.

12.4   REFERENCES

Adriano, D. C. and H. E. Doner. 1982. Bromine, chlorine, and fluorine, In: *Methods of Soil Analysis, Part 2,* 2nd ed., Agronomy No. 9, A. L. Page, R. H. Miller, and D. R. Keeney (eds.). Madison, WI: American Society of Agronomy, 461–462.

Bolton, J. 1971. The chloride balance in a fertilizer experiment on sandy soil. *J. Sci. Food Agri.* 22:292–294.

Bruton, Lowell G. 1971. Known addition ion selective electrode technique for simultaneously determining fluoride and chloride in calcium halophosphate. *Anal. Chem.* 43:479–581.

Carson, P. L. 1980. Recommended nitrate-nitrogen tests, In: *Recommended Chemical Test Procedures for the North Central Region (revised),* North Central Regional Publication No. 221, W. C. Dahnke (ed.). Fargo, ND: North Dakota State University, 12–13.

Christensen, N. W., R. G. Taylor, T. L. Jackson, and A. L. Mitchell. 1981. Chloride effects on water potentials and yield of wheat infected with take-all root rot. *Agron. J.* 73:1053–1058.

Fixen, P. E., G. W. Buchenau, R. H. Gelderman, T. E. Schumacher, J. R. Gerwing, F. A. Cholick, and A. G. Farber. 1986a. Influence of soil and applied chloride on several wheat parameters. *Agron. J.* 78:736–740.

Fixen, P. E., R. H. Gelderman, J. R. Gerwing, and F. A. Cholick. 1986b. Response of spring wheat, barley, and oats to chloride in potassium chloride fertilizers. *Agron. J.* 78:664–668.

Fixen, P. E., R. H. Gelderman, J. R. Gerwing, and A. G. Farber. 1987. Calibration and implementation of a soil Cl test. *J. Fert. Issues* 4:91–97.

Frankenberger, Jr., W. T., M. A. Tbatabai, D. C. Adriano, and H. E. Doner. 1996. Bromine, chlorine, and fluorine, In: *Methods of Soil Analysis, Part 3, Chemical Methods.* SSSA Book Series No. 5, D. L. Sparks (ed.). Madison, WI: Soil Science Society of America, 833–867.

Gaines, T. P., M. A. Parker, and G. J. Gascho. 1984. Automated determination of chloride in soil and plant tissue by sodium nitrate. *Agron. J.* 76:371–374.

Gelderman, R. H., J. L. Denning, and R. J. Goos. 1998. Chlorides, In: *Recommended Chemical Soil Test Procedures for the North Central Region,* North Central Regional Research Publication No. 221 (revised), J. R. Brown (ed.). Columbia, MO. Missouri Agricultural Experiment Station SB 1001, University of Missouri, 49–52.

Glass, A. D. M. 1989. *Plant Nutrition: An Introduction to Current Concepts.* Boston, MA: Jones and Barlett Publishers.

Hipp, A. W. and G. W. Langdale. 1971. Use of a solid-state chloride electrode for chloride determinations in soil extractions. *Commun. Soil Sci. Plant Anal.* 2:237–240.

Johnson, G. V. and P. E. Fixen. 1990. Testing soils for sulphur, boron, molybdenum, and chlorine. In: *Soil Testing and Plant Analysis,* 3rd ed., SSSA Book Series No. 3, R. L. Wasterman (ed.). Madison, WI: Soil Science Society of America, 265–273.

Mosko, J. A. 1984. Automated determination of inorganic anions in water by ion chromotography. *Anal. Chem.* 56:629–633.

Parker, M. A., G. J. Gascho, and T. P. Gaines. 1983. Chloride toxicity of soybeans grown in Atlantic Coast flatwoods soils. *Agron. J.* 75:439–443.

Richards, L. A. (ed.). 1954. *Diagnosis and Improvement of Saline and Alkali Soils,* U.S. Department of Agriculture Handbook No. 60. Washington, DC: Government Printing Office.

Sims, J. T. and G. V. Johnson. 1991. Micronutrient soil tests, In: *Micronutrients in Agriculture,* 2nd ed., SSSA Book Series No. 4, J. J. Mortvedt (ed.). Madison, WI. Soil Science Society of America, 427–476.

Small, H., T. S. Stevens, and W. A. Bauman, 1975. Novel ion exchange chromatography method using conductance detection. *Anal. Chem.* 47:1801–1809.

Timm, C. A., R. J. Goos, A. E. Johnson, F. J. Sobolik, and R. W. Stack. 1986. Effect of potassium fertilizers on malting barley infected with common root rot. *Agron. J.* 78:197–200.

Watson, M. E. and R. A. Isaac. 1990. Analytical instruments for soil and plant analysis, In: *Soil Testing and Plant Analysis,* 3rd ed., SSSA Book Series No. 3, R. L. Westerman (ed.). Madison, WI. Soil Science Society of America, 691–740.

13 Organic and Humic Matter

Soil organic matter exists in two forms: crop and microbial residues, which depending upon soil temperature and moisture, are continuously undergoing decomposition; and humus, an end product of organic matter decomposition which is very stable and contributes to soil structural stability and the water-holding and cation exchange capacities of the soil (Baldock and Skjemstad, 1999). Crop and microbial residues, upon decomposition, are the source for a number of essential plant nutrient elements, such as nitrogen (N), phosphorus (P), and boron (B); while humus impacts the effectiveness of applied soil herbicides (Weber and Peter, 1982; Strek and Weber, 1983; Strek, 1984). The characterization of soil organic matter has been described in detail by Swift (1996).

The commonly used organic matter determination procedures, wet oxidation (Walkley, 1947; Graham, 1948; Nelson and Sommers, 1996; Combs and Nathan, 1998) and loss-on-ignition (Golden, 1987; Ben-Dor and Banin, 1989; Nelson and Sommers, 1996; Combs and Nathan, 1998), do not distinguish between these two forms of organic material in the soil. Therefore, those who use either one of these organic matter determination procedures for predicting N release, may either significantly under or over estimate N contributions to a growing crop.

The loss-on-ignition procedure for organic matter determination is still undergoing study and modification (Golden, 1987).

13.1 WET DIGESTION

1. Principle of the Method

1.1. The total soil organic matter is routinely estimated by measuring organic carbon (C) content. The procedure is described by Mebius (1960).

1.2. The method described is a wet-oxidation procedure using potassium dichromate ($K_2Cr_2O_7$) with external heat and back titration to measure the amount of unreacted dichromate. This method and other methods are thoroughly discussed by Hesse (1971), Jackson (1958), and Allison (1965). The procedure is rapid and adapted for routine analysis in a soil testing laboratory. It is primarily used to determine the organic matter of mineral soils.

2. Range and Sensitivity

2.1. The method is useful for soils containing very low organic C to as high as 12% organic C with a sensitivity of about 0.2 to 0.5% organic C.

3. Interferences

3.1. Soils containing large quantities of chloride (Cl), manganous (Mn^{2+}) and ferrous (Fe^{2+}) ions will give high results. The Cl interference can be eliminated by adding silver sulfate (Ag_2SO_4) to the Oxidizing Reagent (see 6.1). No known procedure is available to compensate for the other interferences.
3.2. The presence of $CaCO_3$ up to 50% causes no interferences.
3.3. This procedure is not recommended for high organic matter content soils or organic soils (see 2.1).

4. Precision and Accuracy

4.1. The method is an incomplete digestion and a correction factor must be applied. The correction factor used is 1.15 (Allison, 1965).
4.2. Repeated analyses should give results with a coefficient of variability of no greater than 1 to 4%.
4.3. Soil samples must be thoroughly ground and mixed before subsampling because heterogeneity is a serious problem in organic matter distribution within samples.

TITRATION PROCEDURE

5. Apparatus

5.1. No. 10 (2-mm opening) sieve.
5.2. 500-mL Erlenmeyer flasks with ground glass tops.
5.3. Reflux condenser apparatus and hot plate.
5.4. Titration apparatus and burette.
5.5. Glassware and pipettes for dispensing and preparing reagents.
5.6. Analytical balance.

6. Reagents (use reagent grade chemicals and pure water)

6.1. *Potassium Dichromate Reagent* (0.267 N): Weigh 13.072 g potassium dichromate ($K_2Cr_2O_7$) into a 1,000-mL volumetric flask, add 400 mL water to dissolve, and then add 550 mL conc. sulfuric acid (H_2SO_4). Let it cool and bring to volume with water.

6.2. *Mohr's Salt Solution* (0.2 M): Weigh 78.390 g ferrous ammonium sulfate [$Fe(NH_4)_2(SO_4)_2 \cdot 6H_2O$] into a 1,000-mL volumetric flask, add 500 mL water to dissolve, and then add 50 mL conc. H_2SO_4. Let it cool and bring to volume with water. Prepare fresh for each use.

6.3. *Indicator Solution:* Weigh 200 mg n-phenylanthranilic acid into a 1,000-mL volumetric flask containing 0.2% sodium carbonate (Na_2CO_3) solution.

7. Procedure

7.1. *Determination:* Weigh 0.1 to 0.5 g (depending on estimated organic content) of air-dry, <10-mesh (2-mm) soil into a 500-mL Erlenmeyer flask and add 15 mL 0.267 N Potassium Dichromate Reagent (see 6.1). Connect the flask to the reflux condenser and boil for 30 minutes. Let cool. Wash down the condenser and flush with

Organic and Humic Matter

pure water. Add 3 drops of the Indicator Solution (see 6.3). Titrate with Mohr's Salt Solution (see 6.2) at room temperature. As the end point is approached, add a few more drops of the Indicator Solution. The color change is from violet to bright green.

8. Calibration and Standards

8.1. Reagents 6.1 and 6.2 should be carefully prepared.
8.2. A blank analysis is carried through the procedure where no soil is added.

9. Calculation

9.1. % organic carbon (C) = {[meq $K_2Cr_2O_7$ - meq $FeSO_4$) × 0.3] / grams soil} × 1.15]
9.2. % organic matter = % organic carbon × 1.724.

10. Effects of Storage

10.1. Air dry soil may be stored many months in closed containers without affecting the organic matter content of the soil.

11. Interpretation

11.1. The interpretation of the organic matter content of a soil will depend on the use of this parameter. Organic matter content can be used to evaluate nitrogen (N) availability and herbicide effectiveness reaction.

COLORIMETRIC PROCEDURE

5. Apparatus

5.1. No. 10 (2-mm opening) sieve.
5.2. 1.5-cm^3 scoop, volumetric.
5.3. 200-mL test tube.
5.4. Delivery burette, or 20-mL automatic pipette.
5.5. UV-VIS spectrophotometer for measuring absorbance (A) at 645 nm (red filter).
5.6. Analytical balance.
5.7. Volumetric flasks and pipettes as required for preparation of reagents and standard solutions.

6. Reagents (use reagent grade chemicals and pure water)

6.1. *Sodium Dichromate Reagent* (0.67 M): Weigh 4.000 g sodium dichromate ($Na_2Cr_2O_7$) into a 1,000-mL volumetric flask and bring to volume with water.
6.2. *Technical Grade Sulfuric Acid* (H_2SO_4).

7. Procedure

7.1. *Determination:* Weigh 2.0 g or volume scoop 1.5-cm^3 (see 5.2) of air-dry <10-mesh (2-mm) soil into a 200-mL test tube (see 5.3). Under a hood, add 20 mL 0.67 M Sodium Dichromate Reagent (see 6.1) and then add 20 mL H_2SO_4 (see 6.2). Mix thoroughly but slowly in order to keep the soil and digestion mixture off the sides of the flask and allow it to cool at least 40 minutes. After cooling, add 100 mL water, mix, and allow it to stand at least eight hours. An aliquot of the clarified solution is transferred to a spectrophotometer vial using a syringe pipette. Measure absorbance (A) at 645 nm (see 5.5).

8. Calibration and Standards

8.1. *Standard curve:* Establish with several soils that have an adequate range of organic matter contents. The percentage of organic matter is determined by a standard method (see *Titration Procedure*). Absorbance values are determined for each known soil organic matter and a curve is constructed by plotting the percentage of organic matter versus absorbance, including a reference sample with daily runs of the method aids in verifying equivalent conditions between the standard curve and daily runs.

9. Effects of Storage

9.1. Soils may be stored in an air-dry condition with no effect on the percentage of organic matter.

13.2 LOSS-ON-IGNITION (LOI)

1. Principle of the Method

1.1. Total soil organic matter is estimated by loss-on-ignition (LOI). The procedure was initially described by Davies (1974), and the method described here is given by Ben-Dor and Banin (1989).

1.2. A soil sample is dried at 105°C (221°F) and then ashed at 400°C (752°F). The loss in weight between 105°C (221°F) and 400°C (752°F) constitutes the organic matter content. The results obtained compare favorably with those obtained by the dichromate wet-oxidation method and by carbon analyzers (Gallardo and Saavedra, 1987; Golden, 1987; Ben-Dor and Banin, 1989; Lowther et al., 1990; Schulte et al., 1991). Others have used different ashing temperatures ranging from 360°C (680°F) (Schulte et al., 1991) to 600°C (1,112°F) (Ben-Dor and Banin, 1989; Gallardo and Saavedra, 1987).

2. Range and Sensitivity

2.1. The method is useful for soils containing low to very high organic matter contents with a sensitivity of about 0.2 to 0.5% organic matter.

3. Interferences

3.1. The method is generally considered not suitable for organic matter determination for calcareous soils. The presence of calcium carbonate ($CaCO_3$) may interfere.

Organic and Humic Matter

4. Precision and Accuracy

4.1. Consistent analytical results are obtainable under a range of sample sizes, ashing vessels, and ashing temperatures and length of ashing time. However, the mineral composition of the soil may be a factor in the determination and may require more than one calibration curve (Schulte et al., 1991). In addition, soil horizons may be another factor affecting LOI results (David, 1988).

4.2. An automated system for determining organic matter content by LOI has been described by Storer (1984).

4.3. Repeated analyses should give results with a coefficient of variability of no greater than 1 to 4%.

5. Apparatus

5.1. No. 10 (2-mm opening) sieve.
5.2. 50-mL beaker or other suitable ashing vessel.
5.3. Drying oven.
5.4. Muffle furnace.
5.5. Analytical balance (0.01 g sensitivity).

6. Procedure

6.1. *Determination:* Weigh 5.00 to 10.00 g sieved (see 5.1) soil into an ashing vessel (see 5.2). Place the ashing vessel containing soil into the drying oven (see 5.3) set at 105°C (221°F) and dry for four hours. Remove the ashing vessel from the drying oven and place in a dry atmosphere. Once cool, weigh to the nearest 0.01 g. Place the ashing vessel plus soil into a muffle furnace (see 5.4) and bring the temperature to 400°C (752°F). Ash in the furnace at 400°C (752°F) for four hours. Remove the ashing vessel from the muffle furnace, let cool in a dry atmosphere and weigh to the nearest 0.01 g.

7. Calibration and Standards

7.1 The percent organic matter in the soil is determined by the formula:

$$\% \ OM = [(W_{105} - W_{400}) \times 100]/W_{105}$$

where: W_{105} is weight of soil at 105°C and W_{400} is weight of soil at 400°C.

7.2. A standard curve may be established with several soils that have a range of organic matter contents encompassing that in the unknowns. The percentage of organic matter in the standards will have been determined by other methods. More than one calibration curve may be required for varying soil types (David, 1988; Schulte et al., 1991).

8. Effects of Storage

8.1. Soils may be stored in an air-dry condition with no effect on the percentage of organic matter.

13.3 HUMIC MATTER BY 0.2 N SODIUM HYDROXIDE EXTRACTION

1. Principle of the Method

1.1. This extraction method is designed to determine the sodium hydroxide (NaOH) soluble humic matter which consists of humic and fulvic acids. These components comprise approximately 85 to 90% of the soil humus and are responsible for the cation and anion exchange properties exhibited by the soil organic fraction.

1.2. This method is based on the concept that humic matter compounds are soluble in dilute alkali solutions (Mortensen, 1965; Levesque and Schnitzer, 1967; Hayes et al., 1975; Tucker, 1992). Acidic organic compounds are converted to ions with the subsequent formation of a physical solution of these ions in water (Mehlich, 1984). The reaction of a dilute alkali with the humic matter results in a colored solution indicating the soluble humic matter content within the soil. The color varies from shades of brown to black, depending on the type of soil from which the sample originates. Colorimetric determination of the humic matter content of soils by this method is based upon the color intensity of the solution following extraction with a dilute alkali extractant. The alkali used in the method is NaOH, which serves as the humic acid solvent (Tucker, 1992). DTPA aids in the dispersion of some of the large molecular calcium (Ca)-humate compounds, and ethanol aids in the solubility of hydrophobic lipid components of soil organic matter. Calibration data was generated from a standard humic matter source (see 4.5).

1.3. This method was designed to accomplish two major objectives: (i) to estimate the chemically reactive portion of the soil organic fraction for better prediction of herbicide rate requirements and (ii) to remove chromium (Cr) from the effluent of municipal waste systems. Experimental evidence has shown that this method can be used to predict herbicide rates (Strek and Weber, 1983; Strek, 1984).

2. Range and Sensitivity

2.1. Up to 10% of the humic matter content of soils can be determined by this method. Higher levels could be determined with a wider extraction ratio (Mehlich, 1984). The method as described will encompass a majority of mineral soils. Saturation of the method is encountered on the organic and mineral organic soils where total organic matter is high. However, in some organic soils, the humic matter content is low even though the percentage of combustible organic content may be in excess of 90%.

2.2. The sensitivity of this method would depend on the quality and homogeneity of the field sample.

3. Apparatus.

3.1. No. 10 (2-mm opening) sieve.
3.2. 1.0-cm^3 (volumetric) soil measure and Teflon-coated leveling rod.
3.3. 55-mL polystyrene extraction vials (35 mm D × 75 mm H).
3.4. Automatic dispenser for extractant, 20-mL capacity.

Organic and Humic Matter   181

3.5. Diluter-dispenser, 5:35-mL capacity.
3.6. Analytical balance.
3.7. UV-VIS spectrophotometer suitable for measuring in the 650 nm range. Spectrophotometers equipped with moveable fiber optic probes can be used to read samples directly from diluted sample vials.

4. Reagents (use reagent grade chemicals and pure water)

4.1. *Extracting Reagent* (0.2N NaOH - 0.0032M DTPA - 2% alcohol): Using a 4,000-mL volumetric flask, add about 1,000 mL water, 32 g sodium hydroxide (NaOH), and dissolve. Then add 16 mL diethylenetriaminepentaacetic acid, pentasodium salt (DTPA) and 80 mL ethanol (C_2H_5OH). Make to volume with water and mix thoroughly. Larger volumes of extractant can be prepared, depending on the numbers of samples to be analyzed.
4.2. *Standard Humic Acid* (Aldrich Chemical Co., 940 W. St. Paul Ave., Milwaukee, WI 53233).

5. Procedure

5.1. *Standard Humic Matter Calibration:* Dry the Humic Acid Standard (see 4.2) at 105°C (221°F) for approximately four hours. Loss on ignition at 550°C (1022°F) shows that this humic matter standard contains 87% organic matter. For calibration, weigh 0.115 g Standard Humic Acid (0.10.87 = 0.115) and place into a 55-cm^3 polystyrene vial (see 3.3). Add 20 mL Extracting Solution (see 4.1) with sufficient force to mix the sample.

Allow the sample to set for one hour and then add an additional 20 mL Extracting Reagent (see 4.1) with sufficient force to mix well. The two 20-mL portions of Extracting Reagent (see 4.1) are added separately to enhance dissolution of the humic matter. Let the sample set overnight (16–18 hours minimum), then pipette 5 mL of the supernatant and 35 mL water into 55-mL polystyrene vials (see 3.3) (*Note:* Caution should be taken not to pipette colloidal precipitation from the bottom of vials.)

The final dilution of the sample is a 1:8 ratio (5 mL sample + 35 mL water) which is required at this extraction ratio to get within the UV-VIS spectrophotometer reading range. Set the spectrophotometer at 100% T with 5 mL Extracting Reagent (see 4.1) and 35 mL water. Read the standard at 650 nm. When a Brinkman probe spectrophotometer with a 2-cm light path is used, the standard humic acid standard should read 10% T. This equates to 10 g 100 cm^{-3} humic matter equivalent. A standard curve can be developed by sequential 1:1 dilutions of the 10% humic acid standard. To develop the factor for converting the spectrophotometer reading to g HM 100 cm^{-3}, convert %T to absorbance, then divide g HM 100 cm^{-3} by the absorbance. Assuming linearity of the standard, the ratio of g HM absorbance^{-1} should be a constant.

If a larger volume of humic acid standard is required for calibration, multiple quantities of standard humic acid and extractant can be used.
5.2. *Soil Sample Extraction and Analysis:* Measure 1 cm^3 soil (screened 2-mm) (see 3.1) into 55-mL polystyrene vial (see 3.3) and add 20 mL Extracting Reagent (see 4.1) with sufficient force to mix well. After one hour, add another 20 mL Extracting

Reagent (see 4.1) with mixing force and allow the samples to set overnight. In addition to allowing adequate reaction time of humic matter with the Extracting Reagent (see 4.1) setting allows soil particles to settle out, leaving a clear supernatant. Transfer 5 mL of undisturbed supernatant and 35 mL water into a 55-mL polystyrene vial. Set the UV-VIS spectrophotometer to read 100% T with a blank (5 mL extractant + 35 mL water). Read the samples at 650 nm and record the %T. A check sample whose humic matter content has been previously determined should be analyzed routinely with unknown samples. Samples which exceed 10% HM can be diluted with water and the appropriate dilution factor employed.

5.3. *Calculations:* The humic matter (HM) content of a soil can be determined from a standard curve or by converting %T to absorbance (Abs) and multiplying by the factor developed in the calibration procedure (see 5.5). For this method the factor is 10, therefore Abs × 10 = gm HM equiv 100 cm^{-3} of soil. If the percentage of HM on a weight basis is desired, divide HM (g 100 cm^{-3}) by the WV (weight/volume in g cm^{-3}) of each soil. For specific values in the development of this procedure (see 5.4).

5.4. *Calibration Procedure:* The values shown below were developed to determine HM up to 10%, using an extraction ratio of 1:40 [1 cm^3 soil + 40 mL Extracting Solution (see 5.1)], with 0.115 g humic acid standard (87% organic matter).

HM Equiv,[a] g 100 cm^{-1}	Abs	HM Equiv/Abs	Factor[c].
10.000	—	—	
5.000	—	—	
2.500	—	—	
1.250[b]	1.000	1.25	10
0.625	0.509	1.23	
0.313	0.206	1.21	
0.156	0.131	1.19	
0.078	0.061	1.28	
0.036	0.027	<u>1.33</u>	
		Avg 1.25 × 8 = 10	

a. The standard HA sample is diluted sequentially (1:1) with water for the development of the standard curve.

b. The standard HA sample is diluted 1:8 (5 cm^3 sample extract + 35 mL water). Read at 650 nm = 10% T or 1.0% absorbance. Unknown samples can be diluted in the same manner.

c. The factor is determined by taking the average of HM per absorbance multiplied by 8 (d.f.)

13.4 REFERENCES

Allison, L. E. 1965. Organic carbon, In: *Methods of Soil Analysis, Part 2, Chemical and Microbiological Properties,* Agronomy No. 9, C. A. Black (ed.). Madison, WI: American Society of Agronomy, 1367–1378.

Baldock, J. A. and J. O. Skjemstad. 1999. Soil organic carbon/Soil organic matter, In: *Soil Analysis: An Interpretation Manual* K. I. Peverill, L. A. Sparrow, and D. J. Reuter (eds.). Collingwood, Australia: CSIRO Publishing, 159–170.

Ben-Dor, E., and A. Banin. 1989. Determination of organic matter content in arid-zone soils using a simple "loss-on-ignition" method. *Commun. Soil Sci. Plant Anal.* 20(15–16): 1675–1695.

Combs, S. M. and M. V. Nathan. 1998. Soil organic matter, In: *Recommended Chemical Soil Test Procedures for the North Central Region*, North Central Regional Research Publication No. 221 (revised) J. R. Brown (ed.). Columbia, MO: Missouri Agricultural Experiment Station SB 1001, University of Missouri, 53–58.

David, M. B. 1988. Use of loss-on-ignition to assess soil organic carbon in forest soils. *Commun. Soil Sci. Plant Anal.* 19:1593–1599.

Davies, B. E. 1974. Loss-on-ignition as an estimate of soil organic matter. *Soil Sci. Soc. Amer. Proc.* 38:150–151.

Gallardo, J. F. and J. Saavedra. 1987. Soil organic matter determination. *Commun. Soil Sci. Plant Anal.* 18:699–707.

Golden, A. 1987. Reassessing the use of loss-on-ignition for estimating organic matter content in noncalcareous soils. *Commun. Soil Sci. Plant Anal.* 18:1111–1116.

Graham, E. R. 1948. Determination of soil organic matter by means of a photoelectric colorimeter. *Soil Sci.* 65:181–183.

Hayes, M. H. B., R. S. Swift, R. E. Wardle, and J. K. Brown. 1975. Humic materials from an organic soil: A comparison of extractants and properties of extracts. *Geoderma* 13:231–245.

Hesse, P. R. 1971. *Soil Chemical Analysis*. New York: Chemical Publishing Co.,

Jackson, M. L. 1958. *Soil Chemical Analysis*. Englewood Cliffs, NJ: Prentice-Hall, Inc.

Levesque, M. and M. Schnitzer. 1967. The extraction of soil organic matter by base and chelating resin. *Can. J. Soil Sci.* 47:76–78.

Lowther, J. R., P. J. Smethurst, J. C. Carlyle, and E. K. S. Mabiar. 1990. Methods for determining organic carbon in Podzolic sands. *Commun. Soil Sci. Plant Anal.* 21:457–470.

Mebius, L. J. 1960. A rapid method for the determination of organic carbon in soil. *Anal. Chem. Acta.* 22:120–124.

Mehlich, A. 1984. Photometric determination of humic matter in soils. A proposed method. *Commun. Soil Sci. Plant Anal.* 15:1417–1422.

Mortensen, J. L. 1965. Partial extraction of organic matter, In: *Methods of Soil Analysis, Part 2, Chemical and Microbiological Properties*, Agronomy No. 9, C. A. Black (ed.). Madison, WI: American Society of Agronomy, 1401–1408.

Nelson, D. W. and L. E. Sommers. 1996. Total carbon, organic carbon, and organic matter, In: *Methods of Soil Analysis, Part 3, Chemical Methods*. SSSA Book Series No. 5. D. L. Sparks (ed.). Madison, WI: American Society of Agronomy, 961–1010.

Schulte, E. E., C. Kaufmann, and J. B. Peter. 1991. The influence of sample size and heating time on soil weight loss-on-ignition. *Commun. Soil Sci. Plant Anal.* 22:159–168.

Storer, D. A. 1984. A simple high sample volume ashing procedure for determination of soil organic matter. *Commun. Soil Sci. Plant Anal.* 15:759–772.

Strek, H. J. 1984. Improved herbicide rate recommendations using soil and herbicide property measurements. Ph.D. dissertation. Crop Science Department, North Carolina State University, Raleigh, NC.

Strek, H. J. and J. B. Weber. 1983. Update of soil testing and herbicide rate recommendations. *Proc. South. Weed Sci. Soc.* 36:398–403.

Swift, R. S. 1996. Organic matter characterization In: *Methods of Soil Analysis, Part 3, Chemical Methods* SSSA Book Series No. 5. D. L. Sparks (ed.). Madison, WI: American Society of Agronomy, 1011–1069.

Tucker, M. R. 1992. Colorimetric determination of humic matter with 0.2 NaOH extraction, In: *Reference Soil and Media Diagnostic Procedures for the Southern Region of the United States,* Southern Cooperative Series Bulletin No. 374, S. J. Donohue (ed.). Blacksburg, VA: Virginia Agricultural Experiment Station, 43–45.

Walkley, Allen. 1947. A critical examination of a rapid method for determination of organic carbon in soils: Effect of variation in digestion conditions and of inorganic soil constituents. *Soil Sci.* 63:251–257.

Weber, J. B. and C. John Peter. 1982. Absorption, bioactivity, and evaluation of soil tests for Alachlor, Acetochlor and Metolochlor. *Weed Sci.* 30:14–20.

14 Organic Soils and Soilless Growth Media Analysis

14.1 DETERMINATION OF pH, SOLUBLE SALTS, NITRATE-NITROGEN, CHLORIDE, PHOSPHORUS, POTASSIUM, CALCIUM, MAGNESIUM, SODIUM, BORON, COPPER, IRON, MANGANESE, AND ZINC IN GREENHOUSE GROWTH MEDIA (SOILLESS MIXES) BY WATER SATURATION EXTRACTION

1. Principle of the Method

1.1. Growth media (soilless mixes) used for the production of plants in greenhouses provide relatively low nutrient-holding capacity. The soil solution is the primary source of nutrient elements for plant growth. A water saturation extract of the growth media, therefore, gives a good indication of the available nutrient element status. The medium is saturated with water without preliminary preparation. This procedure eliminates possible segregation of mix components and insures analysis of the growth medium as the grower is actually using it. Soluble salt and nutrient element concentrations in the water saturation extract are related to the moisture-holding capacity of the growth medium. This process eliminates the need to consider bulk density as a factor in the analysis procedure. One set of guidelines can be used with all soilless mixes (Warncke and Krauskopf, 1983; Berghage et al., 1987; Kidder, 1992; Warncke, 1986, 1990, 1995, 1998; Whipker et al., 1994).

1.2. Water-saturation extraction for measuring the salt content of soil was adopted by the U.S. Salinity Laboratory (1954). During the 1960s, Geraldson (1957, 1970) used the saturated-soil extract approach to determine the nutrient element *intensity and balance* in the sandy soils of Florida. Lucas et al. (1972) studied the saturated soil extract method for analyzing greenhouse growth media (soilless mixes) and found it provided more meaningful results and was more advantageous than the Spurway (1949) method. Sonneveld and van den Ende (1971) and Sonneveld et al. (1974) describe procedures for extraction using 1:2 and 1:1.5 volume extracts for growing media, respectively.

1.3. Water-soluble levels of the key micronutrients in prepared growth media (soilless mixes) are quite low. Zinc (Zn) and manganese (Mn) concentrations in water saturation extracts of growth media rarely exceed 0.8 mg L^{-1}, and iron (Fe) concentrations rarely exceed 4.0 mg L^{-1}. Hence, it is difficult to distinguish between deficient and adequate levels. In peat- and bark-based growth media, the micronutrients are complexed by organic compounds (Verloo, 1980). In evaluating 15 extractants, Berghage et al. (1987) found that extractable levels of Cu, Fe, Mn, and Zn could be increased greatly by using weak solutions of various salts, acids, or chelates in the saturating solution with the saturation-extract procedure. A 0.005 M DTPA reagent was found to increase extractable micronutrient levels most consistently while having only a minor effect on the other key test parameters: total soluble salts and extractable levels of nitrate-nitrogen (NO_3-N), phosphorus (P), potassium (K), calcium (Ca), magnesium (Mg), sodium (Na), boron (B), and chloride (Cl).

2. Range, Sensitivity, and Analysis Methods

2.1. This method is adapted to growth media (soilless mixes and very sandy soils) that have a relatively low nutrient-holding capability. For soils or mixes having appreciable nutrient-holding capacity, this method of analysis reflects less accurately actual nutrient element availability.

2.2. Sensitivity of analysis is dependent upon the instrumentation used. Dilution may be necessary for samples having high soluble salt levels.

2.3. Instruments used include pH meter (Sumner, 1994); conductivity meter (Rhoades, 1996) for the measurement of soluble salt level; spectrophotometry or ion chromatography (Watson and Isaac, 1990; Tabatabai and Frankenberger, 1996) for the determination of NO_3-N, Cl, B, and P or specific ion electrodes (Dahnke, 1971; Mills, 1980; Watson and Isaac, 1990) for NO_3-N and Cl determination; flame emission for the determination of K and Na; atomic absorption spectrophotometry (AAS) (Watson and Isaac, 1990; Wright and Stucznski, 1996) for the determination of Ca, Mg, Cu, Fe, Mn, and Zn; inductively coupled plasma emission spectrometry (ICP-AES) (Soltanpour et al., 1998) for the determination of P, K, Ca, Mg, Na, B, Cu, Fe, Mn, and Zn.

3. Interferences

3.1. pH measurements may be confounded by high-soluble salt levels.
3.2. Soilless mixes containing slow-release fertilizers may give inflated results.
3.3. Interferences relevant to each analytical procedure apply.

4. Precision and Accuracy

4.1. Reproducibility of results is dependent upon wetting the sample just to the point of complete saturation. When properly saturated, pH, soluble salt and nutrient element levels are reproducible.

5. Apparatus

5.1. 600-mL plastic beaker.
5.2. Spatula.
5.3. Filter paper (Whatman No. 1 or equivalent), 11 cm.
5.4. 250-mL vacuum flask.
5.5. Buchner funnel (11 cm) fitted with appropriate rubber stopper to fit vacuum flask (see 5.4).

5.6. Vacuum pump.
5.7. Vial, snap-cap, 100 mL.
5.8. Conductivity bridge with 0 to 1 million ohm range.
5.9. Conductivity cell, dipping type (cell constant of 1.0).
5.10. Thermometer, 0-100°C (32-212°F).
5.11. pH meter with paired glass and calomel reference electrodes.
5.12. pH meter with expanded scale or specific ion meter.
5.13. Nitrate electrode with paired reference electrode.
5.14. Chloride electrode with paired reference electrode.
5.15. UV-VIS spectrophotometer.
5.16. Flame emission, atomic absorption spectrophotometer (AAS), or inductively coupled plasma emission spectrometer (ICP-AES).
5.17. Ion exchange chromatograph.
5.18. Volumetric flasks and pipettes as required for preparation of reagents and standard solutions.

6. Reagents (use reagent grade chemicals and pure water)

6.1. *Extraction Reagent* (Pure Water): For saturation of samples.

6.2. *0.005 M DTPA Reagent* (for saturation of samples in determination of Fe, Mn, and Zn): For each liter of solution, weigh 1.97 g dry DTPA (diethylenetriaminepentaacetic acid) into a 1,000-mL volumetric flask. Bring to volume with 50°C (122°F) water to facilitate dissolution. Allow it to cool to room temperature and bring to volume with water.

6.3. *Potassium Chloride* (0.01 M): Dissolve 0.7456 g potassium chloride (KCl) in water. Bring up to 1,000 mL with water (see Appendix B).

    A. To determine the cell constant (q), use the 0.01 N KCl (see 6.3) solution at 25°C (77°F) which will have a specific conductance (SC) of 0.0014118 dS m^{-1}.

    B. The cell constant (q) for any commercially prepared conductivity cell can be calculated according to Willard et al. (1968, page 720) by the relationship:

$$K = (1/R)(d/A) = q/R$$

where: K = specific conductance, A = electrode area, d = plate spacement, and R = resistance in ohms cm^{-1}. In the case of 0.01 N KCl (see 7.1), the cell constant q = 0.0014118 (in dS m^{-1}) x R (in mS cm^{-1}). R = 708.32 ohm if the cell has electrodes 1 cm^2 in an area spaced 1 cm apart (Note: mhos = 1 per ohm).

    C. Some conductivity instruments read in specific conductance (SC) expressed in mhos x 10^{-5} as well as resistance (ohms). Before accepting the mhos x 10^{-5} dial readings, the cell constant should be determined and the mhos x 10^{-5} dial readings substantiated as being correct for the cell constant used.

6.4. *Buffer Solutions:* pH 4.0 and 7.0 buffers for standardizing pH meter (see Appendix B).

6.5. *Nitrate-Nitrogen Standard* (1,000 mg NO$_3$-N L^{-1}): Weigh 7.218 g potassium nitrate (KNO$_3$) into a 1,000-mL volumetric flask and bring to volume with water.

Prepare standards containing 1, 5, 10, 50, 100, and 200 mg NO_3-N L^{-1} by diluting appropriate aliquots of the 1,000 mg NO_3-N L^{-1} standard with water (see Appendix B).

6.6. *Chloride Standard* (1,000 mg L^{-1}): Weigh 2.103 g potassium chloride (KCl) into a 1,000-mL volumetric flask and bring to volume with water. Prepare standards containing 1, 5, 10, 50, 100, and 200 mg Cl L^{-1} Cl by diluting appropriate aliquots of the 1,000 mg Cl L^{-1} standard with water (see Appendix B).

6.7. *Phosphorus (P), Potassium (K), Calcium (Ca), Magnesium (Mg), and Sodium (Na), Boron (B), Copper (Cu), Manganese (Mn), Iron (Fe), and Zinc (Zn) Standards:* Use commerically prepared 1,000 mg L^{-1} standards or see Appendix B for standard preparation procedures.

7. Procedure

Extraction with Pure Water

7.1. Fill a 600-mL beaker about 2/3-full with the growth media. Gradually add pure water (see 6.1) while mixing until the sample is just saturated. At saturation the sample will flow slightly when the container is tipped and can be easily stirred with a spatula. Depending on the growth media composition, the saturated sample may glisten as it reflects light. After mixing, allow the sample to equilibrate for one hour and then check the following criteria to ensure saturation. The saturated sample should have no appreciable free water on the surface, nor should it have stiffened. Adjust as necessary by adding either additional media or pure water. Then allow to equilibrate for an additional half hour.

7.2. Determine pH of the saturated sample by carefully inserting the electrodes directly into the slurry. Transfer the saturated sample to a Buchner funnel (see 5.5) lined with a filter paper (see 5.3). Be sure there is good contact between the filter paper and funnel surface by eliminating entrapped air. Insert a funnel stopper into the neck of the vacuum flask (see 5.4), apply the vacuum and collect the extract. Transfer the extract to a snap-cap vial.

7.3. Check the temperature of the extract and adjust the temperature dial on the conductivity bridge. Rinse the electrode, then dip the conductivity cell into the extract. Determine the electrical conductance of the extract and record in mS cm^{-1} (or dS m^{-1}).

7.4. After establishing the standard curve, determine the NO_3-N content with a nitrate electrode. Record millivolt reading on an expanded-scale pH meter or specific ion meter and compare with the standard curve plotted on semilogarithmic graph paper (Anon., 1989) or calculate from slope of the line. Nitrate-N can also be determined by spectrophotometry or by ion chromatography (see 2.3).

7.5. After establishing the standard curve, determine the Cl content with a chloride electrode. Record the millivolt reading from an expanded-scale pH meter or specific ion meter and compare with the standard curve plotted on semilogarithmic graph paper or calculate from slope of the line. Chloride can also be determined by ion chromatography (see 2.3).

7.6. Determine the P content in an aliquot of the extract by either a spectrophotometric procedure or by ICP-AES (see 2.3).

7.7. Determine the K, Ca, Mg, and Na content in an aliquot of the extract by either flame emission and AAS or by ICP-AES (see 2.3).

Extraction with 0.005M DTPA to Improve Extraction for the Micronutrients

7.8. Place 400 cm³ of growth media in a 600-mL beaker. Add 100 mL 0.005 M DTPA Reagent (see 6.2). Mix, gradually adding pure water to bring the media just to the point of saturation. See 7.1 for saturation criteria and equilibration time.

7.9. When DTPA is used in the saturation procedure, the media pH must be determined on a separate sample of the media. Use one part media by volume and two parts water by volume.

7.10. Extract the media solution as described in 7.1.

7.11. See 7.3 to 7.7 for the determination of total soluble salts, NO_3, Cl, P and K, Ca, Mg, and Na, respectively.

7.12. Determine the Cu, Fe, Mn, and Zn content in an aliquot of the DTPA extract by AAS or the B, Cu, Fe, Mn, and Zn content by ICP-AES (see 2.3).

8. Calibration and Standards

8.1. *Working Standards:* Working standards should be prepared as indicated in Section 6.7. If the element concentrations are outside the range of the instrument or standards, prepare a suitable dilution. Dilute only as necessary to minimize magnification of the error introduced by diluting.

8.2. *Calibration:* Procedures vary with instrument technique and type of instrument. Take every precaution to ensure that the proper procedures and manufacturer recommendations are followed in the operation and calibration of the instruments used.

9. Calculation

9.1. Report soluble salt levels as mS cm^{-1} or dS m^{-1}. The electrical conductance (dS m^{-1}) can be converted to mg L^{-1} by multiplying by 640, theoretically. However, multiplying by 700 seems to provide a better working conversion factor (Anon., 1989).

9.2. Report results for NO_3-N, P, K, Ca, and Mg as mg L^{-1} of extract.

9.3. Determine nutrient element balance by calculating the percent of total soluble salts for each nutrient element as follows:

$$\% \text{ element} = \frac{(\text{element concentration})(100)}{\text{total soluble salt concentration}} = \frac{(\text{mg L}^{-1})(100)}{(\text{mg L}^{-1})}$$

10. Effects of Storage

10.1. This procedure permits the extraction of moist samples just as they come from greenhouses. Drying of samples is unnecessary and undesirable as changes in nutrient element status may occur. Storage of plant growth media in either dry or moist state will influence primarily the soluble NO_3-N level. When necessary, store at about 4°C (39°F).

11. Interpretation

11.1. Optimum pH, soluble salt and nutrient levels vary with the greenhouse crop being grown and management practices. The following general guidelines can be used in making preliminary judgement of the results (Kidder, 1992; Warncke, 1988, 1990, 1995; Whipker et al., 1994).

Analysis	Level of Acceptance				
	Low	Acceptable	Optimum	Very High	High
Soluble salt, dS m^{-1}	0–0.75	0.75–2.0	2.0–3.5	3.5–5.0	5.0+
Nitrate-N mg L^{-1}	0.39	40–99	100–199	200–299	300+
Phosphorus mg L^{-1}	0–2	3–5	6–10	11–18	19+
Potassium mg L^{-1}	0–59	69–149	150–249	250–349	350+
Calcium mg L^{-1}	0–79	80–199	200+		
Magnesium mg L^{-1}	0–29	30–69	70+		

11.2. In the desired nutrient balance, the total soluble salts are comprised of the following percentages of elements: 8% NO_3-N, 12% K, 15% Ca, and 5% Mg. If Cl and Na are determined, the percentage of each should be less than 10%. If ammonium-nitrogen (NH_4-N) is determined, its concentration should be less than 3%.

11.3. The following general interpretation guidelines should be used in the DTPA extraction procedure (see 7.2) for B, Cu, Fe, Mn, and Zn. Specific desirable levels will vary with the crop being grown.

Micronutrient	Generally Adequate Range mg L^{-1}
Boron (B)	0.7 — 2.5
Copper (Cu)	0.5 — 1.5
Iron (Fe)	15 — 40
Manganese (Mn)	5 — 30
Zinc (Zn)	5 — 30

14.2 REFERENCES

Anon. 1989. *Standard Methods for the Examination of Water and Wastewater,* 17th ed. Washington, DC: American Public Health Association,

Berghage, R. D., D. M. Krauskopf, D. D. Warncke, and I. Widders. 1987. Micronutrient testing of plant growth media: Extractant identification and evaluation. *Commun. Soil Sci. Plant Anal.* 18:1089–1110.

Dahnke, W. C. 1971. Use of the nitrate specific ion electrode in soil testing. *Commun. Soil Sci. Plant Anal.* 2:73–84.

Geraldson, C. M. 1957. Soil soluble salts—determination of and association with plant growth. *Proc. Florida State Hort. Soc.* 71:121–127.

Geraldson, C. M. 1970. Intensity and balance concept as an approach to optimal vegetable production. *Commun. Soil Sci. Plant Anal.* 1:187–196.

Kidder, G. 1992. Determination of pH, soluble salts, nitrate, phosphorus, potassium, calcium, magnesium, sodium, and chloride in potting media (nonsoil mixes) by saturation extraction, In: *Reference Soil and Media Diagnostic Procedures for the Southern Region of the United States,* Southern Cooperative Series Bulletin No. 374, S. J. Donohue (ed.). Blacksburg, VA: Virginia Agricultural Experiment Station, 28–32.

Lucas, R. E., P. E. Rieke, and E. C. Doll. 1972. Soil saturated extract method for determining plant nutrient levels in peats and other soil mixes. *4th International Peat Congress* 3:221–230.

Mills, H. A. 1980. Nitrogen specific ion electrodes for soil, plant, and water analysis. *J. Assoc. Off. Anal. Chem.* 63:797–801.

Rhoades, J. D. 1996. Salinity: Electrical conductivity and total dissolved solids, In: *Methods of Soil Analysis, Part 3, Chemical Method,* SSSA Book Series No. 5, D. L. Sparks (ed.). Madison, WI: Soil Science Society of America, 417–435.

Soltanpour, P. N., C. W. Johnson, S. M. Workman, J. B. Jones, Jr., and R. O. Miller. 1998. Advances in ICP emission and ICP mass spectrometry. *Adv. Agron.* 64:28–113.

Sonneveld, C. and J. van den Ende. 1971. Soil analysis by a 1:2 volume extract. *Plant Soil* 35:505–516.

Sonneveld, C., J. van den Ende, and P. A. van Dijk. 1974. Analysis of growing media by means of a 1 1/2 volume extract. *Commun. Soil Sci. Plant Anal.* 5:183–202.

Spurway, C. H. and K. Lawton. 1949. *Soil Testing,* Michigan Agricultural Experiment Station Bulletin 132. East Lansing, MI: Michigan Agricultural Experiment Station.

Sumner, M. E. 1994. Measurement of soil pH: Problems and solutions. *Commun. Soil Sci. Plant Anal.* 25:859–879.

Tabatabai, M. A. and W. T. Frankenberger, Jr. 1996. Liquid chromatography, In: *Methods of Soil Analysis, Part 3, Chemical Method.* SSSA Book Series No. 5. D. L. Sparks (ed.). Madison, WI: Soil Science Society of America, 161–245.

U.S. Salinity Laboratory Staff. 1954. *Diagnosis and Improvement of Saline and Alkali Soils, USDA Agricultural Handbook No. 60.* Washington, DC: U.S. Government Printing Office.

Verloo, M. G. 1980. Peat as a natural complexing agent for trace elements. *Acta Hort.* 99:51–56.

Warncke, D. D. 1986. Analyzing greenhouse growth media by the saturation extraction method. *Hort. Sci.* 21:223–225.

Warncke, D. D. 1990. Testing artificial growth media and interpreting the results, In: *Soil Testing and Plant Analysis,* 3rd ed. SSSA Book Series No. 3. R. L. Westerman (ed.). Madison, WI: Soil Science Society of America, 337–357.

Warncke, D. D. 1995. Recommended test procedures for greenhouse growth media, In: *Recommended Soil Testing Procedures for the Northeastern United States,* 2nd ed., Northeast Regional Publication Bulletin 493, J. Thomas Sims and Ann Wolf (eds.), Newark, DE: Agricultural Experiment Station, University of Delaware.

Warncke, D. D. 1998. Greenhouse root media, In: *Recommended Chemical Soil Test Procedures for the North Central Region,* North Central Publication No. 221 (revised), J. R. Brown (ed.). Columbia, MO: Missouri Agricultural Experiment Station SB 1001, University of Missouri, 61–64.

Warncke, D. D. and D. M. Krauskopf. 1983. *Greenhouse Growth Media: Testing and Nutritional Guidelines,* Michigan State University Cooperative Extension Bulletin E-1736. East Lansing, Michigan Agricultural Experiment Station.

Watson, M. E., and R. A. Isaac. 1990. Analytical instrumentation for soil and plant analysis, In: *Soil Testing and Plant Analysis,* 3rd ed., SSSA Book Series No. 3. R. L. Westerman. (ed.). Madison, WI: Soil Science Society of America, 691–740.

Whipker, B. E., T. Kirk, and P. A. Hammer. 1994. Industry root media analysis results: Useful in determining greenhouse nutrition problems and educational opportunities. *Commun. Soil Sci. Plant Anal.* 25:1455–1469.

Willard, H. H., L. L. Merritt, Jr., and J. A. Dean. 1968. *Instrumental Methods of Analysis,* 4th ed., Princeton, NJ: D. Van Nostrand Co., Inc.

Wright, R. J. and T. I. Stucznski. 1996. Atomic absorption and flame emission spectrometry, In: *Methods of Soil Analysis, Part 3, Chemical Method,* SSSA Book Series No. 5, D. L. Sparks (ed.). Madison, WI: Soil Science Society of America, 65–90.

15 Quality Assurance Plans for Agricultural Testing Laboratories

15.1 INTRODUCTION

Although it is the intent of every agricultural testing laboratory, regardless of size, to provide reliable testing results to its clientele, each measurement made by the laboratory is associated with some degree of uncertainty. This uncertainty is introduced into every process step of an analysis, because the analysis itself is designed to estimate unknown values. To meet the goal of reliable testing results, a laboratory must have some way of proving its credibility. This section will discuss one method of providing credibility: a quality assurance (QA) program.

Here the term "process" refers to a discrete part of a measurement methodology. "Analysis" refers to the result of the overall methodology. Often, agricultural measurements consist of several process steps within an analysis. For example, for soil testing, the processes of sample preparation, sample extraction, analytical measurement, data reduction, and final report might describe an analysis. Each process step introduces some uncertainty into the final data produced from the analysis.

15.2 QUALITY ASSURANCE IN THE LABORATORY

Providing credibility through a QA program is something that at first may seem to be counterproductive. That is, the more testing a laboratory does to prove that unknown measurements are correct, the fewer unknown samples the laboratory can complete in a given time period.

In fact, once the QA process is fully implemented throughout the laboratory, the laboratory may actually be analyzing more unknowns. The reason for this apparent contradiction is twofold. First, the accuracy and precision will be known for the system and, therefore, can be demonstrated to the client with fewer QA samples. Second, when the system of testing is within statistical control (i.e., data from the QA program meet preselected statistics), fewer unknowns will have to be rerun. Thus, the efficiency of the process is increased and the total cost for completing unknown samples decreases. (See the section on statistical control.)

The ability to reproduce a value time and again (precision) or the ability to attain a known result (accuracy) indicates that the testing process operates within known uncertainty levels. If accuracy and precision levels are proven to be within the goals of both the client and the internal laboratory guidelines throughout a sufficiently long time period and testing range, the process is much more likely to continue to produce quality data in the future. Only a few tests to demonstrate continued accuracy and precision of the process are required.

Uncertainty can be introduced into the measurement process from many sources. Technical staff work within the uncertainty of the process. When the uncertainty of the process is known, digressions from the normal level of uncertainty can be detected easily and appropriate changes made before additional testing, of inferior quality, continues. Technicians can detect inferior results immediately, spending valuable time more wisely (recalibrating, etc.) rather than being told about the error later by the manager and having to repeat previous testing. The reduced workload also minimizes chemical waste, which, in turn, keeps operating costs low.

15.2.1 Knowing How a Laboratory Operates

Producing sample analyses is the real purpose of a laboratory. Knowing how the process really works, a leader (manager) can identify possible problem areas within a process. An effective leader will remove any stumbling blocks to quality with which the technical staff are confronted. The technical staff will work within the process, while the leader works on the process itself (Deming, 1986).

Figure 15.1 describes the chain of command in a typical laboratory facility. However, the diagram does not explain how the work is actually accomplished, nor does the figure indicate who does the work. Instead, the figure implies that the director initiates all processes. The implication that the technical staff contribute little to the laboratory other than labor is inherent in such a diagram (Rummler and Brache, 1995). Yet, it is the training and expertise of the technical staff that contribute directly to the success of the laboratory. For this reason, these diagrams are of little benefit in improving the internal processes of the laboratory, and do not contribute directly to the QA process of the laboratory.

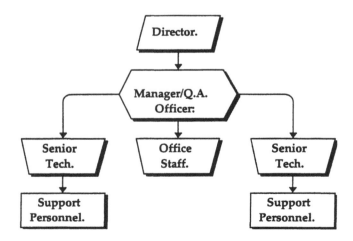

FIGURE 15.1 Chain of command for an analytical laboratory.

The internal processes or work paths can better be described for QA improvement if diagrams such as Figure 15.2 are developed. The information presented in Figure 15.2 is a partial diagram of the sample process, showing a simplified pathway for nitrates. The main contributors to the process are listed across the page in columns, while the work completed by the contributors is listed in the various shaped boxes. Each horizontal line connecting boxes indicates that one contributor transfers his/her work to another contributor. In other words, one is acting as a supplier and the other as a customer.

It is here that the QA management perspective can be promoted. If the customer has to do additional work (inspections, data entry, etc.) to bring the "product" of the supplier to standards, a process change at the supplier side may be needed. Additionally, if the supplier is unaware of the needs of the customer, further QA problems can be introduced. Quality begins with the supplier. It is the leader's responsibility to address the process, on both supplier and customer sides, to remove any impediments to quality.

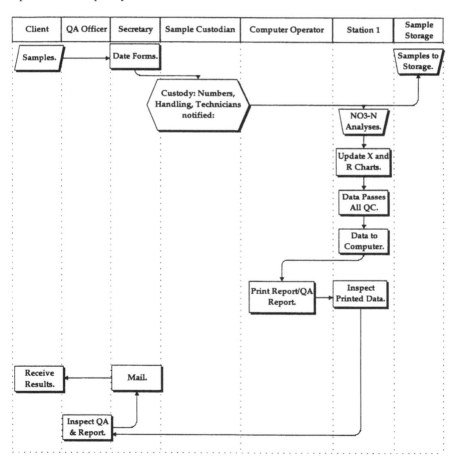

FIGURE 15.2 Internal process or work paths in an analytical laboratory.

It is often useful to have the technical staff create the initial version of these diagrams and to discuss the results in staff meetings. Since the technical staff works within each of these processes, who is better able to explain the process as it actually exists? When similar diagrams were first drafted at the University of Florida's Institute of Food and Agricultural Sciences Analytical Research Laboratory, it was found that a rerouting of the work path eliminated some unneeded steps and reduced the overall workload by 25 percent for one technician. This change, initiated after the new process had been tested, had no measurable effect on the quality of data originating from the process.

Once the process has been diagrammed to the satisfaction of all contributors the standard operating procedures (SOP) within a process can be identified and prepared. The diagrams must take into account all sample handling, data transfer, and paperwork within a process to ensure that the manager can address all possible points within the process needing QA activity. The written SOP text contains only those parts of the process that need further explanation not provided in the diagram (see the section on documentation on pages 202–205).

15.3 STATISTICAL CONTROL

The uncertainty of a process and its major contributing components must be quantified. Any QA program is wasted effort if the statistical control of a process has not been documented. The QA program provides the "user of a product or a service the assurance that it meets defined standards of quality with a stated level of confidence" (Taylor, 1987). When the uncertainty of a process has been documented, and these data meet the predetermined limits of the QA program, the process is said to be within statistical control.

Another type of diagram, often called a "fishbone diagram," is useful for breaking a process into the parts that directly contribute to the final quality of the result (Ishikawa, 1976). Figure 15.3 shows a fishbone diagram (nonexhaustive) that illustrates possible sources of uncertainty for a nitrate colorimetric process. Each main component contributing to uncertainty can further be broken down and indicated as additional branching within the main component. The manager can then make choices concerning further testing of a process component if the overall uncertainty of the process is not within acceptable limits.

If the process is found to be within current limits, these diagrams can help the manager to decide where to focus attention to further reduce current limits. Additionally, these diagrams are useful for training new technicians since the diagrams point out the components of the process that contribute to uncertainty.

15.3.1 STATISTICAL MEASUREMENTS

The measurements needed to insure that the process is within acceptable limits can take many forms, some of which are discussed below. No subset of these measurements is best for all processes. The manager must choose the method or

Quality Assurance Plans for Agricultural Testing Laboratories

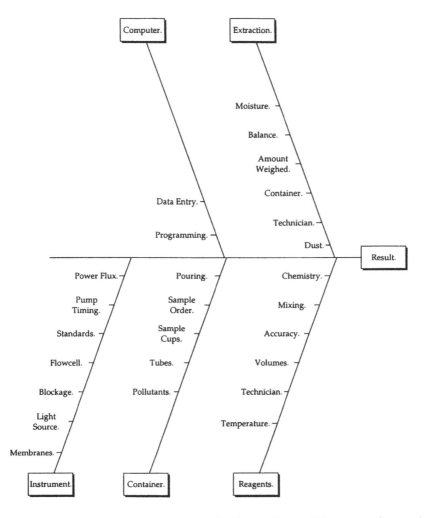

FIGURE 15.3 Fishbone diagram (nonexhaustive) illustrating possible sources of uncertainty for a nitrate colorimetric process.

methods that are best for the particular process within a specific laboratory. The following list is not exhaustive but provides a wide spectrum of possibilities.

External standards are issued by some source outside of the laboratory. The National Institute of Standards and Technology, for example, specifies the mean measurement with some level of confidence about that mean. Thus, the external standard provides an unbiased estimate of the laboratory's ability to determine accurately a value for the measurement.

Accuracy is the agreement of the measured value from the laboratory and the expected value supplied by the source of the external standard. Use of the external standard is considered unbiased because the laboratory itself does not specify the expected value.

An internal standard is a reference material that has been prepared within the laboratory. While of great value to the QA program, this class of standards should be considered biased because the laboratory itself specifies both observed and expected values.

Most external standards are costly, and frequent use increases the operating costs of the laboratory. Internal standards are usually much less expensive. Thus, if a relationship between external and internal standard can be firmly established, frequent use of the internal standard with occasional use of the external standard can reduce costs while still providing acceptable QA. The manager should explore a mixture of both external and internal standards to ensure accuracy. Acceptance of this mixture may be directly dependent on the needs of the client.

15.3.2 REPLICATES OF UNKNOWNS

Repeated measurements of an unknown sample give an estimate of the precision of the process. Precision of the process is the level of agreement among independent measurements using the specified conditions of the process. Use of this QA technique can provide information on the stability of the process to reproduce the values within a certain level of uncertainty. It has the added advantage of generating such results directly from the client's samples, not from standards that may have a differing matrix from the client's samples. In addition, the material is supplied by the client, usually at no extra cost to the laboratory.

As with external and internal solutions, the manager must strike a balance between the number of replicates and the production of completed samples by the laboratory. When the process is in statistical control, replicates can be held to a minimum of, for example, one duplicate in every 20 unknown samples. This number should increase, if the process is working, within about five times the value of the process's lower limit of detection.

15.3.3 SPIKED UNKNOWNS

This method of determining quality measures the percentage of recovery of an internal standard when it is mixed with an unknown sample. While often used in QA programs, this method may introduce dissimilar matrix problems into the process. Additional errors can be introduced in the measurement and selection of the quantity and level of the spiking solution. Taylor (1989) advised that spiking should not be used for calibration and validation when other means are available. For this reason, spiking should not be frequently conducted on samples for normal agricultural testing. While this procedure may be of some benefit to the QA program, it can also introduce serious problems.

15.4 CONTROL CHARTS

Control charts can be the backbone of a good QA program. Data collected on a frequent basis from these charts can be combined to produce process audits by the manager and summary QA documents for the client. Some of the many types of control charts—those that can be most easily used by the technical staff—are presented here.

X charts can be used for both external and internal standards. Figure 15.4 shows the daily values for an internal standard of 10 mg nitrate-nitrogen (NO_3-N) L^{-1} with time used at the Extension Soil Testing Laboratory, University of Florida. Daily readings such as these demonstrate the time stability of the process. Drifting or out-of-control values can be detected at the work station and recalibration initiated before large numbers of unknown samples must be rerun. If the QA program is truly of highest concern within all parts of the laboratory, the goal will not be to produce acceptable or "within specification" results. Rather, the goal will be to produce what can be achieved in the given process.

Figure 15.4 also shows both upper and lower warning limits (UWL and LWL) and control limits (UCL and LCL). These limits are found by calculating the standard deviation of a portion of the data (a minimum of seven points is recommended) and multiplying by 2 for the WL and 3 for the CL. The UWL and LWL lines will include about 95% of the data, while the UCL and LCL lines will include 99%. Rules for deciding about the statistical control of the process are discussed later in this chapter.

A similar chart using the mean of four or more analyses, called the X-bar chart, can also be constructed. However, this chart does require many more additional analyses, and for this reason it is not frequently used in agricultural testing.

The R chart, alternately called a range chart, is a useful method of displaying the range (highest-lowest) of replicated samples. The data displayed in Figure 15.5 are from duplicate analyses. The mean R is calculated on a minimum of 8 values, but preferably with 15 values (Taylor, 1987). The UWL is calculated by multiplying the

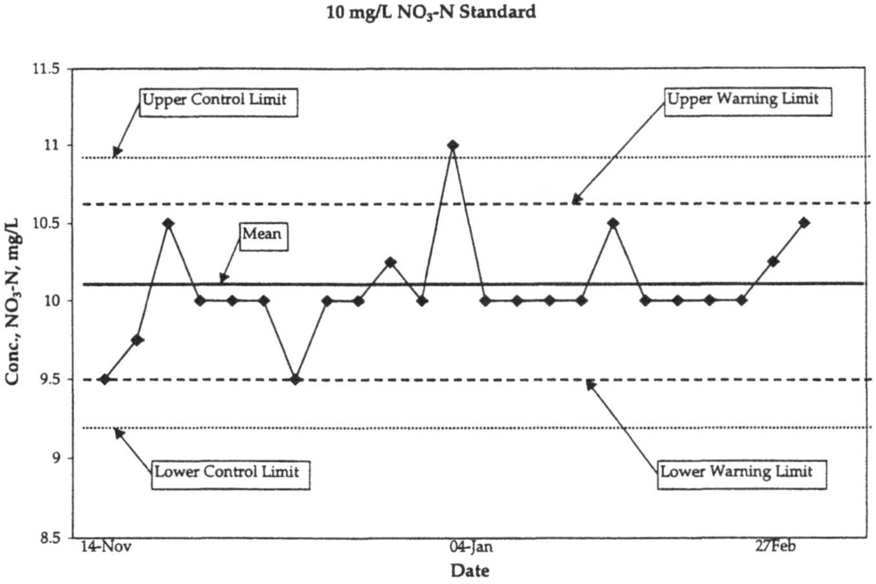

FIGURE 15.4 *X chart* showing the daily values for an internal standard.

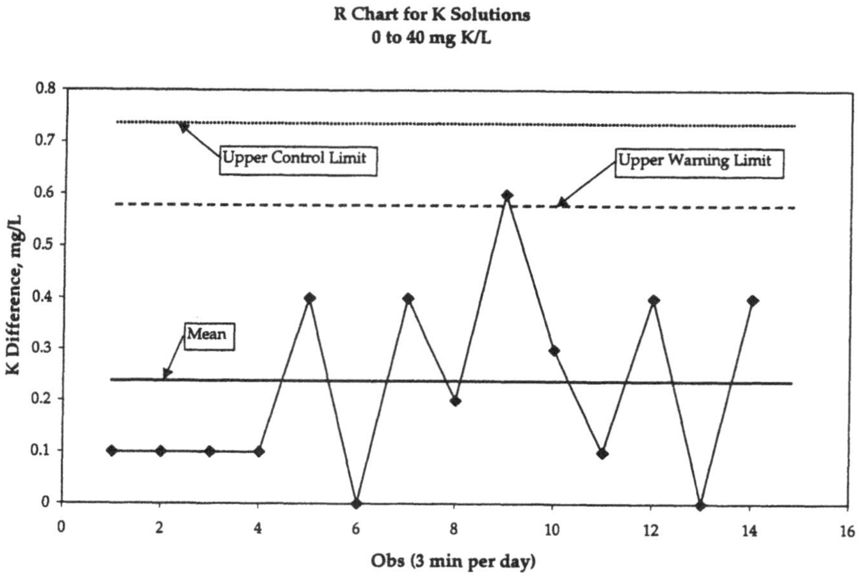

R Chart for K Solutions
0 to 40 mg K/L

FIGURE 15.5 *R chart* (or *range chart*) displaying the range (highest-lowest) of replicated samples.

standard deviation of the data by 2.512, while the UCL is found by multiplying the standard deviation by 3.267.

The R chart will not reflect some process-oriented problems, such as deteriorating instrument calibration. For this reason, the R chart should be used in conjunction with an X chart. Since the R chart is developed by replicating unknown samples, this chart accurately indicates the precision of the process.

The R chart can be used with replicates of other than duplicate size (Taylor, 1987). The R chart should be used for replicate samples that are within a reasonable range of values. Replicated samples that are of widely differing reported values should be used with caution. Selection of the range must be based upon the process. A series of tests is needed for the selection.

Another option, the range performance chart, allows a wide range of readings (Figure 15.6). The chart does not reflect the precision of the system chronologically. This chart was made based upon data that were collected within four concentration groups. Calculations are the same as for the R chart, but mean R values and limits are plotted and a straight line drawn through respective points.

15.4.1 MAKING DECISIONS WITH THE CHARTS

Daily use of the charts can be the first line of defense against poor quality data. In using the charts, the technician should follow a few general rules:

1. If two measurements fall within the warning (UWL or LWL) and control lines (UCL or LCL), reject all data that lie between the most recent reading

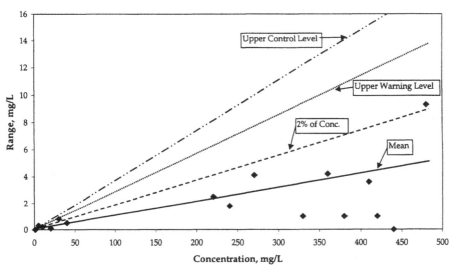

FIGURE 15.6 *Range performance chart* that allows for a wide range of readings to be graphed.

and the last control sample that is known to be in control (reading within the UWL and LWL). Take corrective action, and reread the identified samples.

2. If one measurement falls outside of the control lines (UCL or LCL), reject all data from the most recent sample and the last control sample that is known to be in control. This rule applies to the NO_3-N reading that exceeded the UCL (Figure 15.4). Samples completed since the last in-control standard (10.5 mg NO_3-N L^{-1}) should be rejected. The process should be recalibrated and all rejected samples retested.

3. Identify trends in the data (more than seven points in rough alignment), and take corrective action before control samples fall outside the warning lines.

15.5 BLIND STUDIES

Two levels of blind studies are possible. The first level is composed of one or more samples of known values (external and/or internal standards). The technician is usually aware that these samples will be used in the QA program. However, the technician does not know the expected analytical results.

The second level, the so-called double-blind study, is similar to the first system in every respect, except that the technician cannot distinguish the QA samples from other unknown samples. For example, the QA officer enters a set of external/internal samples into the laboratory using an assumed name. None of the laboratory personnel are aware of the nature of these samples and the set receives the same attention as unknown samples.

The double-blind system is the preferred method of collecting data about the health of the QA process, since it removes any preconceived bias present with the first system. When the process is considered in statistical control, such studies (either blind or double-blind) can be conducted weekly with good results. If the process is not in control, blind studies should not be conducted. Rather, efforts should be focused on establishing control.

15.6 SYSTEMS AUDITS

Systems audits are conducted to ensure that the process is within specified QA objectives. The systems audit should review all of the statistical data generated from the process as well as instrument logs and in-laboratory time for analyses. The discussions should address those parts of the process which (i) are the most difficult, (ii) require the most hand labor, or (iii) are just unpopular. It is the technician's responsibility to detail real or perceived impediments to quality, while the manager must inspect the process for means to remove these impediments.

A systems audit may point out problems with which a technician cannot deal, either due to a lack of training or a lack of specific guidance. Training of both technical and management staff, or even a brief restatement of the goals, is often the solution to process problems.

15.7 DOCUMENTATION

The following parameters will establish the basic paperwork needed for a sound QA program. In all cases, documentation should be used to substantiate the QA program. Paperwork is of little value unless it can be created easily, can be interpreted by those who must use it, and can be read by those summarizing the information contained therein. All documentation must be functional and not merely contribute to overhead costs.

15.7.1 CUSTODY OF SAMPLES

The progress of samples through the laboratory must be accurately documented. Simple forms or computer-generated screens can be used to detail a clear path of sample receipt by the laboratory, showing any work performed within the laboratory and by whom, subsequent sample disposal, and information reported to the client.

Multiple copies of such paperwork should be strongly questioned. For example, it makes little sense for individual technicians to maintain copies. Rather, the central office can maintain the original and a working copy can accompany the samples through the laboratory. In most laboratory information management systems, custody records can be computerized.

15.7.2 Standard Operating Procedures (SOP) and Good Laboratory Procedures (GLP)

These detailed documents (SOP and GLP) are needed to describe the internal functions of the laboratory. They are easily written after the within-laboratory supplier/customer pathways and the fishbone charts are diagrammed. If the process is new to the laboratory, one might further delay until the diagrammed process has been found to be within statistical control. Then, the procedural description will include those techniques that can be used to produce a stable process.

Many formats for SOP and GLP exist. Both the U.S. Environmental Protection Agency (EPA) and the Food and Drug Administration (FDA) have specific guidelines for such documents. State agencies usually will also have specific requirements for acceptable SOP/GLP documentation.

15.7.3 Methodology References

All methodology reference material should be readily available. Many laboratory manuals containing SOP/GLP information also contain the appropriate references. The purpose of such documentation is to show the implementation of the procedure as well as any interferences, etc., which influence the process. Such references are useful both to technicians and managers to ensure that the procedure is being accurately followed. Any changes in the reference methodology should also be documented to avoid confusion and misunderstanding in the future.

15.7.4 Instrument Handling and Maintenance Procedures

Detailed descriptions of proper instrument techniques should be available. A readily available instrument manual may suffice for this guide, but often the details of a specific process may not be addressed in the manufacturer's manual. For this reason, a supplemental guide can be quite helpful, especially if the guide is written with the involvement of the technical staff. A checklist format for daily operations may also help to preclude unsafe procedures or forgetfulness.

Maintenance or operating records on each instrument are of great value since the history of problems and cures is then readily available. An additional benefit of these records is that instrument operating costs and sample volume can be collected with little loss of time. To be accurate, the operating record should include any samples that were reanalyzed due to rejected QC standards by the technical staff. Rejected samples may not be counted in any other way, yet they directly affect the productivity of the laboratory.

Preventive maintenance can also be scheduled to alert the operator of needed care. Early detection and replacement of failing parts can have a positive effect on instrument performance.

15.7.5 QA Expectations and Measured Performance

Each laboratory should record and maintain information regarding each process within the laboratory. A clear statement of the QA objectives for a process can be quite helpful for the client. Such a statement also sets the specifications within which the technical staff should work. The overall management goal must be to do the best job possible. Specifications should be used only as guidelines, not as a listing of what is tolerable.

The measured performance of each process demonstrates that the process is within statistical control. Ideally, the performance should be better than the specifications, except where the specifications force work at or near the limits of some part of the process, such as the lower limit of detection of an instrument.

The level of accuracy and precision demanded of a process should be in part based upon the end use of the analytical results. For environmental research involving suspected toxic materials, the quality should be quite high. Resulting data will be used to judge the harmfulness of the substance, affecting its subsequent or continued use.

In situations where soil or tissue testing is used for nutrient element management decisions, the analytical result is compared to an interpretation scale. This scale was developed on initial laboratory results from field and/or greenhouse plant-production trials. Therefore, while the reliability of the data should be high, the translation of those data using the interpretive instrument arbitrarily sets less critical limits on the data to be interpreted.

For example, a soil-test method that has been calibrated for a specific crop response to added fertility can be interpreted as low, medium, or high (Figure 15.7). This interpretation scale is a traditional step function model; however, similar problems result when regression models are employed. In the latter case, the confidence intervals of the regression serve as the limits of the so-called gray zones.

In this example, an arbitrary error of 15 percent was used to illustrate the concept. This error level is often used by laboratory evaluation schemes. If a laboratory result is found to be within 15 percent of the mean or median of all laboratories, then the result is considered to be acceptable. While the laboratory may be capable of day-to-day variability of <0.1 mg L^{-1} for pure standards, the limit may be widened to approximately 1 mg L^{-1} (for example) due to soil variability, and is further widened due to the interpretation scale imposed upon the interpretation of the resulting soil test for fertilization decisions.

Laboratory variability within a specific interpretation has no effect on the interpretation. Within the indicated gray zones, laboratory variability is critical, moving the interpretation from one interpretation range to the adjacent range, and directly affecting the fertilization recommendation.

It is evident that the quality of soil-test data should be judged within the agronomic or horticultural context from which the soil test derives its usefulness in the first place. This reasonable approach means that laboratory quality control standards for routine agricultural or horticultural testing should not focus on method detection limits, but rather on accuracy and precision within a well-known working range (Figure. 15.7, gray zones). Additionally, control efforts for analyses well above the interpretive zone are also fruitless, adding neither refinement to the interpretation nor improvement to the fertilization recommendation.

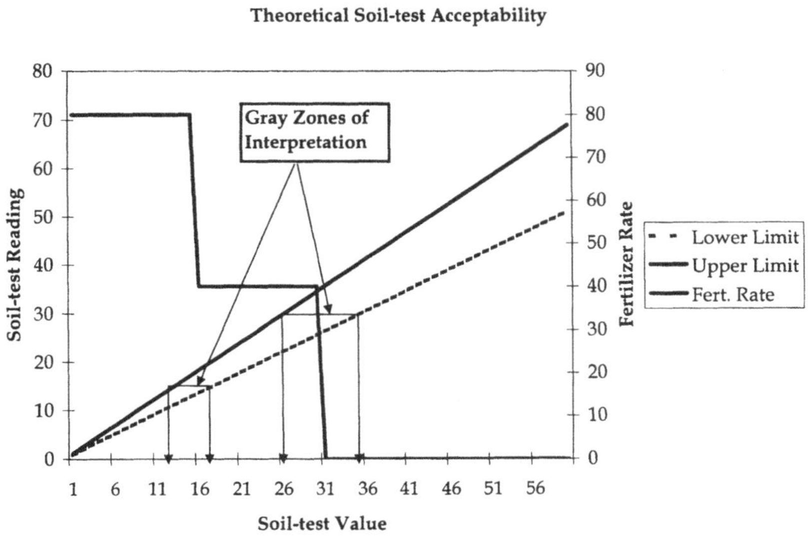

Theoretical Soil-test Acceptability

FIGURE 15.7 Relationship between laboratory variability and resulting interpretation and fertilization recommendation.

15.8 SUMMARY

While a QA program may be viewed as a drain on laboratory resources, such an assessment is not valid wherever an efficient QA program has been adopted. In the soil testing laboratory, a workable QA program has three components:

- QA is always cost-effective when used as a management tool to create a process that is in statistical control.
- The trained technician is the best QA guarantee.
- The laboratory manager modifies the process while the technician works within the process. Both the technician and the manager must work together to remove all impediments to quality.

15.9 REFERENCES

Deming, W. E. 1986. *Out of the Crisis.* Cambridge, MA: Massachusetts Institute of Technology, Center for Advanced Engineering Study.

Ishikawa, K. 1976. *Guide to Quality Control.* Murray Hill Station, NY: Unipub, Asian Productivity Organization.

Taylor, J. K. 1987. *Quality Assurance of Chemical Measurements.* Chelsea, MI: Lewis Publishers.

Rummler, G. A. and A. P. Brache. 1995. *Improving Performance: How to Manage the White Space on the Organization Chart,* 2nd ed. San Francisco, CA: Jossey-Bass Publishers.

16 Methods of Instrumental Analysis

16.1 INTRODUCTION

The elemental concentration in obtained soil extracts can be determined by a number of analytical instrumental procedures with some elements determinable by more than one technique as shown in Table 16.1.

TABLE 16.1
Instrumental methods of analysis by elements and ions.

Elements/Ions	Colorimetric[1]	Flame	Spark Emission	ICP	Atomic Absorption	Specific Ion Electrode
Boron (B)	good	na	good	excel	na	na
Calcium (Ca)	good	fair	good	excel	excel	poor
Copper (Cu)	good	na	good	excel	excel	na
Iron (Fe)	fair	na	good	excel	excel	na
Potassium (K)	poor[2]	excel	excel	excel	good	na
Magnesium (Mg)	fair	fair	excel	excel	excel	na
Manganese (Mn)	good	na	excel	excel	excel	na
Molybdenum (Mo)	good	na	poor	good	good[3]	na
Sodium (Na)	na	excel	excel	excel	good	na
Phosphorus (P)	excel	na	excel	excel	na	na
Zinc (Zn)	good	na	excel	excel	excel	na
Ammonium (NH$_4$)	good	na	na	na	na	good
Chloride (Cl)	good	na	na	na	na	good
Fluorine (F)	na	na	na	na	na	good
Nitrate (NO$_3$)	excel	na	na	na	na	good
Sulfate (SO$_4$)	good[2]	na	na	na	na	na

na = not applicable
excel = excellent (high sensitivity with mininal interference)
good = moderate sensitivity with some interference
fair = reasonable sensitivity but with matrix effects
poor = reasonable sensitivity with significant matrix effects
[1] UV-VIS spectrophotometry
[2] turbidity
[3] flameless atomic absorption spectrophotometry

All but a few of these procedures involve some form of spectrophotometry, the utilization of a specific wavelength to measure either intensity or absorption to determine elemental or ion specie concentration. There are essentially four analytical techniques that have application for such elemental assays: procedures that are still in common use in soil analysis laboratories today; colorimetry (UV-VIS spectrophotometry); turbidity; and emission and atomic absorption spectrophotometry. The trend today is greater analytical sophistication—multi-element computer-controlled analytical instrumentation—resulting in a lessened understanding of the analytical principles involved, and "black box" concepts of instrument calibration, maintenance, and operation. Many technicians today are less knowledgeable about the analytical procedures they are using and more concerned as to which button to push in order to carry forward an analysis.

Preparation of the analyte and suitable adaptation of the method of analysis require an understanding of the principles of operation of the method as well as its requirements and limitations. Adequate testing is usually required before putting a method into use, following procedures such as those that have been adapted by the Association of Official Analytical Chemists (AOAC) (McLain, 1982).

Although the elemental sensitivity of the analytical technique (see Table 16.2) is a significant factor for some procedures and with some elements (particularly the heavy metals, see Chapter 9), the precision of the method is equally important. Normally the elemental content found in a soil extractant is considerably above the detection limit of most analytical procedures with long-term stability of the method over the assay time period being the most important for determining its acceptance and use. This involves both the characteristics and stability of the analytical instrument as well as the quality of reagents used. The positioning of "check samples" or "reference standards" in an array of sequential extractant unknowns is essential in order to monitor the stability of the analytical run (see Chapter 15). For those times when extractants have elemental concentration content(s) which exceed the upper limit of the procedure, this problem needs to be addressed by either diluting the extractant to bring the elemental concentration down to within the analytical range of the procedure, or report the analytical result as being greater than the upper limit.

TABLE 16.2
Comparison of aqueous detection limits (ng L^{-1}) for atomic optical spectrophotometry methods.

Element	FAAL	FAFL	FAE	RFICP	ETA-AA
Aluminum (Al)	100	100	3.0	0.2	0.1
Arsenic (As)	30	100	10,000	40	0.8
Boron (B)	2,500	—	50	5.0	20
Calcium (Ca)	2.0	20	0.1	0.02	0.04
Cadmium (Cd)	1.0	0.001	800	2.0	0.008
Chromium (Cr)	2.0	5.0	2.0	0.3	0.2
Copper (Cu)	4.0	0.5	10	0.1	0.06
Iron (Fe)	4.0	8.0	5.0	0.3	1.0

Lead (Pb)	10	40	50	2.0	0.4
Magnesium (Mg)	3.0	0.1	70	0.05	0.004
Manganese (Mn)	0.8	1.0	1.0	0.06	0.02
Molybdenum (Mo)	30	500	200	0.2	0.3
Nickel (Ni)	5.0	3.0	20	0.4	0.9
Phosphorus (P)	10	10	100	2.0	0.2
Potassium (K)	3.0	—	0.05	—	4.0
Sodium (Na)	0.08	—	0.5	0.2	—
Zinc (Zn)	1.0	0.02	10,000	2.0	0.003

Key: **FAAL** — flame atomic absorption with line source
FAFL — flame atomic fluorescence with line source
FAE — flame atomic emission
RFICP — radiofrequency inductively coupled plasma
ETA-AA — electrothermal atomization-atomic absorption

Accuracy, another essential requirement, is obtained by using "reference materials" (Ihnat, 1993; Quevauviller, 1996) of established content for calibration and monitoring (see Chapter 15 and Appendix B) as well as applying those quality assurance procedures essential for the successful operation of an analytical laboratory (Taylor, 1987). Participation in an established Proficiency Testing Program (see Chapter 1, Section 1.4) (Houba et al., 1996; van Dijk et al., 1996: Wolf and Miller, 1998) can uncover errors in an analytical procedure and provide assistance in correcting them (Miller et al., 1996).

There are two basic textbooks on analytical chemistry (Scoog et al., 1988; Ewing, 1990) that would be helpful in a soil analysis laboratory, four that deal with specific analytes (Anon., 1986; Rump and Krist, 1988) or elements (van Loon, 1991; Smoley, 1992) of agricultural interest, and books and articles that deal specifically with the analysis of soils (Hesse, 1971; Walsh, 1972; Anon, 1986; Watson and Isaac, 1990; Smith; 1991).

Since it is not possible to cover in adequate detail every aspect of each method of analysis suited for the elemental assay of soil extracts, the objective of this chapter is to provide sufficient background information to guide the user in those analysis procedures suitable for soils and in particular, focusing on those instrumental methods used in this handbook.

16.2 UV-VIS SPECTROPHOTOMETRY

UV-VIS spectrophotometry (formerly known as colorimetry), has had a long history of application and use for elemental determination in soil extractants (Piper, 1942; Jackson, 1958). The invention and use of automatic analyzers, such as the Technicon AutoAnalyzer (Coakley, 1981; Smith and Scott, 1983), and more recently the injection flow analysis procedure (Ranger, 1981; Ruzicka and Hanson, 1988), have kept spectrophotometry an important analytical procedure. Procedures for the use of these automated analyzers have been described by Isaac and Jones (1971), Flannery and Markus (1972), Watson and Isaac (1990), and Smith and Scott (1991).

UV-VIS spectrophotometry has good selectivity and sensitivity for many of the elements (or ions) found in soil extractants, such as ammonium (NH_4), boron (B), copper (Cu), iron (Fe), magnesium (Mg), manganese (Mn), molybdenum (Mo), nitrate (NO_3), phosphorus (P), sulfate (SO_4), and zinc (Zn).

A UV-VIS spectrophotometer is a relatively inexpensive instrument that is easy to use. The main components include a light source to produce a beam of monochromatic light, a sample holder or cell, and detector. Today, most spectrophotometers exhibit improved performance because they employ either a a prism or grating in lieu of an interference filter to obtain a specific monochromatic wavelength of light. Paul (1998) discusses past and continuing advances in the field of UV-VIS instruments.

Spectrophotometry is based on the principle of light absorption by a compound. The amount of light absorbed or the precentage of light transmitted is correlated with the presence and concentration of a particular absorbing substance in solution. This relationship is defined by Beer's Law:

$$\log_{10}(PO/P) = abs \quad \text{or} \quad A = abc,$$

where: transmittance [$\log_{10}(PO/P)$], or absorbance (A), is a function of the constance specific (a) of the substance, its thickness (b), and concentration (c) of the relative number of colored ions or molecules in the light path. According to Beer's Law, there is a linear relationship between absorbance (A) and concentration (c) when monochromatic light (wavelength selection is also critical) is used. Lack of linearity invalidates Beer's Law, making the analytical procedure invalid or risky in its application (lack of linearity at the extremes of a linear curve does not invalidate the linear portion of the calibration curve; only that portion that is nonlinear is of questionable value). The useful concentration range is usually two decades, although it may be extended to three for some elements and procedures. This relatively narrow range is a factor that can limit the usefulness of this technique when assaying extractants whose element/ion concentration range is in excess of two to three decades.

Determining the spectral properties of a formed complex, the wavelength of maximum absorption with the least interference, must be carefully done. For example, the wavelength for the determination of P in soil extracts by the molybdenum-blue procedure is 882 nm (Rodriquez et al., 1994). Care is needed to ensure that the sample solution is of the proper pH and that interfering substances are absent or compensation made for their presence. For some elements and solutions, separation of the element(s) of interest may be needed to ensure removal of interfering substances and/or to concentrate the element of interest prior to analysis. Such procedures may be required when determining a low concentration of a micronutrient or trace element in the presence of high concentrations of other elements.

Turbidimetric spectrophotometry is an analytical technique in which the complex formed is a precipitate rather than a true solution; the spectrophotometer is used to determine the density of the suspended precipitate in the solution. Considerable care is required to ensure that a uniform particle size precipitate is formed that will remain in suspension during the measurement of absorption. The two elements of agricultural interest that can be determined by this analyical technique are K and the

Methods of Instrumental Analysis 211

SO_4 anion. The same general conditions must be met as to the properties of the to-be-assayed solution in terms of its pH and elemental composition in order for the precipitate to be successfully formed. The only turbidimetric procedure used to any great extent is for the determination of SO_4-S (see Chapter 11; Singh et al., 1995; Combs et al., 1998).

16.2.1 AUTOMATED UV-VIS SPECTROPHOTOMETERS

It has been the automation of spectrophotometric and turbidimetric procedures that has kept these techniques still in wide use today, the first being the introduction of AutoAnalyzers, and more recently, flow injection analyzers.

In the late 1950s, Technicon Corporation introduced the AutoAnalyzer (AutoAnalyzer is a registered trade name), a continuous flow system of analysis suitable for a wide range of analytical applications for assaying solutions (Coakley, 1981). The AutoAnalyzer was quickly adapted for use in soil extractant analyses (Isaac and Jones, 1971; Flannery and Markus, 1972; Smith and Scott, 1991). Its rapid acceptance, not only for this application, but for a wide range of other analytical applications (Knopp and McKee, 1983), is based on its excellent ability to do repetitive analytical processes with minimal technician assistance at relatively good speed and excellent control, producing high quality analytical results.

The principle of operation is based on the continuous flow of standard solutions and unknown analytes which are intermittently introduced into a flowing stream of reagents, mixed together by constant inversion through glass mixing coils and segmented by air bubbles equally spaced from one another. These air bubbles establish and maintain sample integrity, promote the mixing of reagent and sample streams, and provide a visual check of the stream-flow characteristics for monitoring the behavior of the system.

The AutoAnalyzer is a train of interconnected modules, its flowing stream being directed through tubing from module to module. The basic components are a sampler, proportioning pump, detector, and recorder. Additional components may be required, such as heating baths and dialyzers, to satisfactorily meet the requirements of the analytical procedure.

Since the AutoAnalyzer is a continuous flow system, reference and sample solutions enter the analytical stream and are segmented into discrete liquid increments, or slugs. As the slug is carried through the system, reagents are added and physical manipulations (mixing, extractions, etc.) are continuously taking place. Since the system is sequential, each event takes place one step at a time with each slug being treated as an individual sample. Proper adjustment between sample and wash ratio is needed to adequately separate samples as determined by testing and experience. A calibration curve is prepared by running a series of suitable standards embracing the range expected in the unknowns. Peak height indicates concentration, and peak heights of unknowns are compared to those of the standards to determine concentration of the element in the unknown.

The introduction of the AutoAnalyzer was a revolutionary advancement in analytical chemistry methodology, and its use standardized many colorimetric procedures (Smith and Scott, 1991); however, the AutoAnalyzer is relatively slow and quite wasteful of reagents. Recently flow injection analysis (Ranger, 1981; Ruzicka and Hansen, 1988), a rapid low volume system in which the analytical measurement is not always made at the equilibrium point of the reaction, is an automated method being rapidly adopted. Using micropump tubes and rapid-moving solutions, response time is in seconds and reagent use minimal. Most of the chemistries used with the AutoAnalyzer are applicable to flow injection.

16.3 FLAME EMISSION SPECTROPHOTOMETRY

This method of analysis is based on the properties of ions (excited atoms) deficient in shell electrons which release absorbed energy in the form of electromagnetic radiation (light) when electrons are captured to produce a ground state atom (all electron shells are complete). This phenomenon can be demonstrated by the simple procedure of placing a platinum wire into a gas flame after dipping it into a solution of sodium chloride (NaCl). A bright yellow color is seen, the result of emitted light energy emanating from Na atoms due to shell electron movement. The wavelength of radiation emitted is characteristic of the element and the intensity of the emitted radiation is correlated with concentration. Wavelength tables giving principal emission lines by element can be found in most handbooks of chemistry and/or physics (Dean, 1973) and books on emission spectrophotometry (Ingle and Crouch, 1988; Metcalfe, 1987; Lajunen, 1992).

Emission spectrophotometry has had a long history of application and use in the agricultural field (Mitchell, 1964). The first development in emission spectrophotometry that had a significant impact on soil analysis methodology was the flame photometer (Mavrodineanu, 1970; Isaac and Kerber, 1971) that first appeared in the 1950s, making K and Na easy determinations. The flame photometer is a spectrophotometer in which the light source is replaced by a flame. The analyte (soil extractant) is carried by the air stream into the flame by means of an aerosol generated by a cross-flow nebulizer. The flame can be a gas mixture of acetylene (C_2H_2) with air (most frequently used), oxygen (O_2) or nitrous oxide (N_2O), or argon (Ar) and hydrogen (H) in entrained air.

The alkaline earth elements, Ca and Mg, can be determined by flame emission, but with difficulty. The presence of other ions (elements) in solution causes interferences that must be either removed or compensated for if accurate determinations are to be made. The reason K and Na are so easily determined by flame excitation is due in large part to their relatively low excitation potentials; that is, there is substantial shell electron movement in the relatively low temperature (1,000–1,500°K) flame. By contrast, the extent of electron movement for the elements Ca and Mg in the same flame is considerably less due to their higher excitation potentials. Therefore, relatively less radiant energy is produced.

Another factor which limits the flame emission techniques is self-absorption. Atoms of the same element which are not in an excited state will absorb the emitted radiation when an atom of the same element is going from the excited to ground state. Since in the excitation source the number of atoms in the ground state far exceeds those in the excited state for most elements, absorption by these ground state atoms significantly reduces the emission to be detected and measured. However, it is this property of self-absorption that led to the development of atomic absorption spectrophotometry, an analytical technique that will be described later.

For biological samples, the use of an internal standard, usually lithium (Li), is highly recommended. Multielement capability is possible, although most commercial instruments have but one or two detectors with or without an internal standard channel. Pickett and Koirtyohann (1969) have compared the detection limits for 62 elements determinable by flame emission and atomic absorption spectrophotometry (see Table 16.2). Recent basic texts on emission spectrophotometry are by Metcalfe (1987), Ingle and Crouch, 1988, and Lajunen (1992) as well as articles by Morrison and Risby (1979), Watson and Isaac (1990), and Wright and Stuczynski (1996).

16.4 ATOMIC ABSORPTION SPECTROPHOTOMETRY

Flame atomic absorption spectrophotometry (frequently referred to as AAS) was introduced in the early 1960s, and at the time, it revolutionized elemental analytical chemistry. With AAS, the analyst was able to easily determine a number of important elements (Ca, Mg, Cu, Fe, Mn, and Zn) of agricultural interest whose alternate assay methods were frequently difficult and tedious.

The principle of operation is the reverse of that for emission in that absorption by ground state atoms of emitted radiation from the same element is utilized. Therefore, a source of radiant energy of the element of interest must be provided as well as a means of bringing into the emitted radiation source the sample containing the same element as ground-state atoms whose concentration is to be determined.

The radiant energy source is provided by means of a hollow cathode lamp in which the cathode is made of the element to be determined. The sample in liquid form containing the element of interest is carried into a flame, commonly an acetylene (C_2H_2)-oxygen (O_2) flame, as an aerosol in the same manner as for analysis by flame emission spectrophotometry. The radiant energy from the hollow cathode lamp is passed through the flame and the reduction in emitted radiation (as a result of absorption by the same element as the cathode present in the analyte) determined. The degree of reduction is correlated with the concentration of the element in the analyte.

The essential components of an AAS spectrometer are the hollow cathode lamp and power source, cross-flow nebulizer and spray chamber, burner, monochromator, and detector.

Only one element at a time can be determined, since most AAS spectrophotometers have a single analytical channel. However, this is not to be confused with AAS spectrophotometers that are single- or dual-beam instruments with one beam being the analytical channel and the other the reference or background channel. There is some set-up time required between elemental sets, as the proper hollow cathode lamp must be put into place and the system optically aligned and calibrated. Most AAS spectrophotometers have some type of processor control to assist the analyst in calibrating the spectrometer. However, compared to other techniques that have polychromators, analysis by AAS is relatively slow and tedious, although the relative performance in terms of accuracy and precision is comparable.

Flame AAS is not totally interference free since it requires the use of "releasing agents" and the addition of an excess of a "competing ion" to minimize interference effects. Lanthanum (La) is the most common releasing agent, while K can be added as the competing ion. The addition of chelating agents, such as 8-hydroxyquinoline or the Na salt of ethylenediamine tetraacetic acid (EDTA), is also effective in reducing interfering effects. The normal elemental operational concentration range is between two and three decades, which may pose a problem for wide-ranging element content samples.

Operating conditions and settings for an AAS are provided by the manufacturer, and they are also found in a number of reference texts related to agricultural applications by Metcalfe (1987) and Lajunen (1992) as well as articles by Isaac and Kerber (1971), Morrison and Risby (1979), Watson and Isaac (1990), Sharp (1991), and Wright and Stuczynski (1996).

The common operating mode is flame AAS, with the introduction of the analyte into the flame as a water aerosol, but for the those elements that form volatile hydrides, such as arsenic (As), bismuth (Bi), lead (Pb), selenium (Se) and tin (Sn), the hydride itself can be introduced directly into the flame (Thompson et al., 1978). The advantages are reduced interferences and increased sensitivity at the nanogram level (Ure, 1991). Although the flameless AAS technique offers excellent sensitivity, sample preparation and handling, as well as matrix interferences, can present difficult problems to the analyst.

Although flame and flameless AAS are very useful analytical techniques, AAS instruments are single element analyzers not well suited for handling multielement assays and have limited concentration range. Flameless AAS is best when high sensitivity is required and/or only small-sized samples are available.

16.5 INDUCTIVELY- COUPLED PLASMA EMISSION SPECTROMETRY

A whole new era for emission spectrometry came into existence with the development and commercialization of plasmas as an excitation source (Fassel and Kniseley, 1974), providing a powerful multielement analysis technique with excellent sensitivity and emission stability. Although the DC plasma (Skogerboe et al., 1976) was the first introduced in the early 1970s, it is the inductively-coupled plasma, frequently referred to as ICP, ICAP, or ICP-AES that has been the most useful of the two sources. The very high temperature of the plasma (8,000 to 10,000°K) results in high emissions, minimizing the effects due to self-absorption and other interferences. Calibration curves are usually linear over several decades of concentration (high sensitivity plus little self-absorption which gives a three- to five-decade linear calibration range), making for easy calibration using the two-point technique as well as allowing for wide-ranging elemental concentration-containing extractants. Direct-reading polychromators and sequentially-operated spectrometers with complete or partial computer control provide the analyst with a variety of operating options.

Most elements found in soil extractants, whether in macro, micro, or trace concentrations, can be easily determined by ICP (see Tables 16.1 and 16.2). As with flame emission, the liquid sample is introduced as an aerosol into the plasma, whose design and operating principles have been described by Scott et al. (1974). Samples with divergent elemental concentrations can be easily handled without the need to dilute or concentrate. The prominent ICP emission lines for elements found in soils and soil extractants are identified along with detection limits in the article by Soltanpour et al. (1998). Details for the use of ICP for the analysis of a range of biological materials have been described by a number of authors (Morrison and Risby, 1979; Soltanpour et al., 1996; Dalquist and Knoll, 1978; Munter and Grande, 1981; Jones, 1977). Thompson and Walsh (1983), Boumans (1987), Montasar and Golightly (1987), and Varma (1991) are authors of recent texts on the ICP technique. Today, laboratories engaged in soil analysis are turning to ICP as the analytical technique of choice.

16.5.1 Spectrometer Designs

16.5.1.1 Sequential Spectrometer

As the name implies, one element at a time is assayed by either physically moving the grating or the detector in the spectrometer during the presence of the sample in the plasma. This spectrometer configuration works best for the assay of substances that have varying elemental compositions and the analyst selects those elements to be determined for each substance assayed. However, this spectrometer configuration is not well-suited for the repeated assay of large numbers of samples for the same suite of elements or for the determination of many elements (greater than 10, for example) in a long series of samples, since analysis is slow and a considerable volume of analyte is required. Depending on the number of elements determined and analysis time, calibration monitoring may be required fairly frequently.

16.5.1.2 Simultaneous Spectrometer

In this configuration, the spectrometer is a polychromator, as each element has its own exit slit and detector; therefore, all the elements so installed in the spectrometer are determined simultaneously during the presence of the sample in the plasma. The method is ideally suited for the repetitious assay of samples. Since only those elements with installed exit slits and detectors can be determined with this spectrometer configuration, some manufacturers will provide for the installation of an additional single-
element detecting spectrometer so that any element not present in the installed polychromator array may be determined. Unfortunately, this added feature may not give good performance because of the characteristics of the added spectrometer (poor sensitivity, inadequate dispersion, etc). Another option is to have the plasma placed physically between two spectrometers, one a simultaneous, the other a polychromator. This is an expensive option but workable for situations that would require such versatility when dealing with requirements for both varied and high-volume analytical requirements.

16.5.1.3 Photodiode Array Spectrometer

This is the latest configuration to show promise of being the spectrometer of the future. A photodiode array serves as the detector. Although significant advances have been made with this type of detector, considerable advancements have yet to be made to equal that of current detector systems. With the photodiode array configuration, there is the marked reduction in size of the spectrometer, meaning that the ICP can be a truly a desktop analyzer similar in physical size to most atomic absorption spectrophotometers. In addition, this spectrometer design offers some unique features for doing "approximate" assays and allows the analyst to design a wide range of analytical programs and systems, which are not possible with current sequential or simultaneous spectrometers.

16.5.2 OPERATING CHARACTERISTICS

16.5.2.1 Advantageous Characteristics

- Multielement, high speed, and very sensitive analytical procedure suitable for elemental content determination of aqueous liquids.
- Sensitivity in the $\mu g\ L^{-1}$ [parts per billion (ppb)] range for many elements due to the very high temperature (8,000 to 10,000°K) of the plasma discharge.
- Linear three- to five-decade elemental concentration reading range due to the fact that relatively few stable atoms (which absorb radiation) are present in the plasma, making the plasma itself what is termed "optically thin" (minimal self-absorption).

16.5.2.2 Disadvantages

- Not all elements exhibit the same emission characteristics in all parts of the plasma; for the elements K and Na, for example, maximum emission occurs at the top and along the edge of the plasma discharge, whereas for most of the heavy metals, maximum intensity exists at the base of the plasma discharge; axial (Petterelli, 1995) or total viewing of the plasma discharge and/or variable positioning of the viewing height are possible means to compensate for these effects.
- Samples containing high levels of organic substances are not easily assayed unless specific operating characteristics are used.
- Samples must be free of suspended materials as clogging or impaired flow rates through the nebulizer will significantly affect the analysis result; changing solution viscosity will alter the flow rate as well as the positioning of the sample container and depth of solution if pneumatic lifting is used to bring the analyte into the nebulizer, whereas peristaltic pumping of the analyte into the nebulizer minimizes flow rate effects.

16.5.2.3 Standard Preparation

Calibration standards must conform exactly to the matrix of the assayed samples as well as bracket the concentration range for the determined elements. The plasma discharge (temperature profile) itself will conform to the dominant element (or ion) present which will, in turn, affect the emission characteristcs for all the other elements in the sample.

For primary standards, it is best to use commercially prepared elemental standards [of 1,000 or 10,000 mg L^{-1} (ppm) concentration] rather than relying on standards prepared from laboratory reagents (see Appendix B). When preparing working standards, compatibility and interelemental factors need to be considered. For most multielement determinations, it is not always possible or wise to include in one standard all the elements to be determined; therefore, some system of standard preparation needs to be followed that minimizes interelement effects or compensates for these effects when unknowns contain widely varying concentrations of certain elements (Jones, 1977).

It is good to use standards from known sources (such as NIST, etc.; see Appendix B) available for verification of accuracy. Multielement-containing standards can be obtained from commercial sources and should be used for verification of prepared working standards.

16.5.2.4 Calibration Techniques

The key to success for the assay of any soil extractant is the setting of the zero when using a two-point calibration routine. The zero point will establish the slope of the calibration curve and set the lower detection limit. For best results, the selected "zero standard" is that calibration standard less only the element to be determined. Therefore, the number of standards, their elemental composition, and selection of the "zero standard" are critical decisions. Once the spectrometer is calibrated, its calibration should be verified by running the standards as "unknowns" along with QA samples following the calibration routine, (see Chapter 15).

Accuracy verification should be conducted using standards from known sources. During an analytical run, a QA sample should begin and terminate the assay run, and/or it may be specifically placed following a set number of unknowns (once every 10th unknown, for example) or randomly spaced among the unknowns. Any variance from the known concentration of an element in the QA sample warns of the need to adjust the analytical result or void the analytical run results completely.

To check the calibration curve for linearity and possible interelement interference(s), high concentration [1,000 mg L^{-1} (ppm) and or higher] single element commercial standards should be assayed following calibration while observing the determined concentration of the element itself plus other elements in the suite of elements being determined. Lack of linearity for the calibration element and the "apparent" presence of other elements may indicate the need for evaluation and correction before the assay of unknowns commences.

The use of an internal standard is not commonly recommended for the ICP instrumental technique; however, there are considerable advantages if the elemental concentrations among the unknowns are high and widely varying. An internal standard would partially compensate for changing sample characteristics that would affect the flow rate (nebulization characteristics) of the unknown(s) and when a wide range in elemental concentration exists from one sample unknown to another. In order to obtain the maximum benefit from the use of an internal standard, its emission characteristics should closely match those of the element(s) being determined; for example, lithium (Li) would be a suitable internal standard element for the determination of K and Na, whereas the elements indium (In) or yttrium (Y) would be best suited for many of the so-called "heavy metals." The internal standard must be added to both standards and unknowns at the same concentration. Adding an internal standard, such as Li, to standards and unknowns at a concentration that exceeds that of any element in the unknowns will stabilize the temperature profile of the plasma discharge, which could significantly improve the precision of the analytical data obtained.

For samples having relatively high concentrations of an element, the "memory effect" should be checked in order to ensure that before the next sample emission is integrated, no significant carryover effect is still apparent. In order to minimize such an effect, it may be necessary to have an extended wash time between samples, or minimize the influence of a carry-over effect by grouping high and low content samples together.

It also may be desirable to have the wash liquid between samples similar to the unknowns (for example, if a particular soil extracting reagent is being assayed, use that reagent as the wash solution), or use the next sample, if sufficient quantity is available, as the wash solution. Simple tricks such as these may significantly improve precision as well as the speed of the analytical procedure.

There is a stabilization (warm-up period) time required (usually 20 to 30 minutes) from the time the plasma is formed and the spectrometer is ready for calibration. An ICP instrument should be placed in a room in which there is controlled stable temperature and humidity as well as isolation from other types of laboratory activity. The discharge from the plasma should be directly vented outside.

16.5.2.5 Common Operating Problems

- Irregular nebulizer flow can occur due to changing viscosity of samples, pump tube wear and/or irregular pumping rate, partial plugging of the nebulizer intake capillaries, varying argon flow, or varying back pressure in the nebulizer chamber.
- Optical alignment is essential for high quality performance. This includes correct positioning of the plasma image on the entrance slit of the spectrometer as well as the alignment of the entrance slit with the exit slits within the spectrometer. Alignment procedures vary with the type of instrument and will change with time and temperature.

- Without high-purity argon, formation of the plasma may be difficult, if not impossible, even with a slight change (purity and flow rate) of the argon from the supply tank.
- Electronic irregularities and failures may occur.

16.5.2.6 Important General Points

As with most complex analytical instruments, an ICP should be operated by a well-trained and experienced operator if reliable analytical data is to be obtained. There is no substitute for experience, as most ICP instruments seem to have "characters of their own" which respond to the care taken by the analyst in all its functions.

The analyst should maintain daily operating logs, noting repairs, service conducted (changing pump tubes, replacing the torch, etc.), standardization data, replacement standards, QA samples, etc. If optical intensity measurements are made, these values should be recorded and monitored. The analyst should monitor continuously the established quality control (QC) procedures, making adjustments for drift or changing values of QAs included in the analytical run (see Chapter 15).

Each ICP has its own set of performance characteristics that may vary from the norm as specified by the manufacturer. For example, elemental sensitivity levels (usually identified as two times the standard deviation), saturation concentrations for elements, degree of linearity for calibration curves, level of precision by element, interelement interferences (apparent and real), etc., are all specific characteristics that will vary among instruments. Therefore, tests should be conducted periodically to observe and record such performance characteristics.

Although elemental detection limits are important characteristics for an ICP instrument, performance characteristics, when assaying soil extractants, need to focus primarily on precision and long-term stability. These are the characteristics most highly desirable when selecting an ICP instrument, since the elemental concentration levels in most soil extractants are at levels considerably above detection limits. In fact, linearity of calibration curves between 100 to $>$ 1,000 mg L^{-1} (ppm), the saturation point of detection sensors, and the apparent or real interelement effects among elements are far more important determining characteristics than detection limits.

16.6 ION CHROMATOGRAPHY (IC)

The ion chromatographic (IC) technique is based on the principle of ion separation using an ion exchange resin. The basic principles of the technique are discussed in the books by Ettre (1980) and Ravindranath (1989), its application to the assay of soil extractants by Tabatabai and Basta (1991) and Tabatabai and Frankenberger (1996), and environmental applications by Frankenberger et al. (1990). Although both cations and anions are determinable by IC, the primary use is for anion determination, the anions being fluoride (F$^-$), chloride (Cl$^-$), phosphate (PO$_4^{3-}$)-P, nitrate (NO$_3^-$)-N, and sulfate (SO$_4^{2-}$)-S, which are listed in their elution order from the ion exchange column. The IC method has a number of significant limitations. One limitation is the slowness of a determination, the total time for one chromatogram of the suite of anions listed above being approximately 20 minutes (time for each anion listed in the order above being 2, 3, 3.5, 6, 11, and 16 minutes, respectively). Another is that the presence of certain reagents and substances in the analyte may not be compatible to the ion exchange column and can "poison" it. Although the IC does have excellent performance characteristics, it has limited applications in the assay of soil extractants. Karmarker (1998) describes the advantages of combining IC with sequential flow injection analysis.

16.7 SPECIFIC-ION ELECTRODES

Specific-ion electrodes work on the same principle as that of the glass electrode for measuring H-ion concentration (Fisher, 1984). These electrodes employ liquid or solid ion exchange membranes, solid membranes composed of single crystals, or precipitates compressed into a plug or dispersed in a matrix such as silicone rubber. The instrument components are the specific-ion and reference electrodes and recording meter. The advantages of specific-ion electrodes are their speed of determination and general simplicity of operation. However, they are not without their limitations, such as their lack of sensitivity when the analyte of interest is in low concentrations and interferences. The nitrate (NO_3^-) and chloride (Cl^-) anions in soil extractants are commonly determined by this technique (see Chapters 10 and 12). Talibuddeen (1991) gives a very thorough description of these electrodes, their design and applications, while descriptions of their use in the assay of soil extractants are given by Carlson and Keeney (1971), Mills (1980), Watson and Isaac (1990), and Talibudeen (1991).

Specific-ion electrodes are available for a number of elemental determinations as shown in Table 16.3.

TABLE 16.3
Commonly available ion-selective electrodes.

Ion		Type of Electrode
Direct	**Indirect**	**Membrane**
Cations		
Ammonium		Liquid (in polymer)
Calcium		Liquid (in polymer)
Copper		Solid State (crystal)
Potassium		Solid State (glass)
		Liquid (in polymer)
Sodium		Solid State (glass)
		Solid State (crystal)
		Liquid (in polymer)

Anions

Bromide		Solid State (crystal)
Chloride		Solid State (crystal)
Cyanide		Solid State (crystal)
Fluoride	Aluminum	Solid State (crystal)
Fluoroborate	Boron	Liquid (free flowing)
Iodide		Solid State (crystal)
Nitrate		Liquid (in polymer)
Sulfide		Solid State (crystal)

The principle of operation is based on the development of a potential across the electrode membrane. This is dependent on the concentration of the ion species being measured in the unknown solution and is measured against a constant reference potential with a specific solution, as described by the Nernst Equation:

$$E = E_o - S \log (A),$$

where: E = measured electrode potential, E_o = reference potential (a constant), A = ion concentration in the unknown solution, and S = electrode slope.

Ionic activity coefficients are variable and largely dependent on total ionic strength. If background ionic strength is high and constant, relative to the sensed ion concentration, the activity coefficient is constant and activity is directly proportional to concentration.

An ionic strength adjuster is normally added to standards and samples so that the background ionic strength is high and constant, relative to the variable concentration of the to-be-measured ion.

Many advancements have been made in the design and use of specific-ion electrodes and their application to soil extractant analysis. Their use requires careful adjustment of the assay solution to minimize interferences and ensure reliable analytical results are obtained.

16.8 REFERENCES

Anon. 1986. *The Analysis of Agricultural Materials,* 3rd ed., Reference Book 427. London: Her Majesty's Stationery Office.

Boumans, W. J. M. (ed.). 1987. *Inductively Coupled Plasma Emission Spectroscopy, Part II: Applications and Fundamentals.* New York: John Wiley & Sons.

Carlson, R. M. and R. D. Keeney. 1971. Specific-ion electrodes: Techniques and uses in soil, plant and water analysis, In: *Instrumental Methods for Analysis of Soils and Plant Tissue,* L. M. Walsh (ed.). Madison, WI: Soil Science Society of America, 39–66.

Coakley, W. A. 1981. *Handbook of Automated Analysis-Continuous Flow Techniques.* New York: Marcel Dekker, Inc.

Combs, S., J. Denning, and K.D. Frank. 1998. Sulfate-sulfur, In: *Recommended Chemical Soil Test Procedures for the North Central Region.* North Central Regional Research Publication 221 (revised), J. R. Brown (ed.). Columbia, MO: Missouri Agricultural Experiment Station SB 1001, University of Missouri, 35–40.

Dalquist, R. L., and J. W. Knoll. 1978. Inductively-coupled plasma-atomic emission spectroscopy: Analysis of biological materials and soil for major, trace, and ultra-trace elements. *Appl. Spectrosc.* 32:1–30.

Dean, J. A. 1973. *Lange's Handbook of Chemistry, Section 8–19,* 11th ed. New York: McGraw-Hill.

Ettre, L. S. 1980. Evolution of liquid chromatography: A historical overview, In: *High Performance Liquid Chromatography: Advances and Perspectives.* Vol. 1, C. Horvath (ed.). New York: Academic Press, 1–74.

Ewing, G. W. (ed.). 1990. *Analytical Instrumentation Handbook.* New York: Marcel Dekker, Inc.

Fassel, V. A. and R. N. Kniseley. 1974. Inductively coupled plasmas. *Anal. Chem.* 46:1155A–1164A.

Fisher, J. E. 1984. Measurement of pH. *Amer. Lab.* 16:54–60.

Flannery, R. L. and D. K. Markus. 1972. Determination of phosphorus, potassium, calcium and magnesium simultaneously in North Carolina, ammonium acetate, and Bray P1 soil extracts by AutoAnalyzer, In: *Instrumental Methods for Analysis of Soils and Plant Tissue,* L. M. Walsh (ed.). Madison, WI: Soil Science Society of America, 97–112.

Frankenberger, Jr., W. T., Mehra, H.C., and D. T. Gjerde. 1990. Environmental applications of ion chromatography. *J. Chromatogr.* 504:211–245.

Houba, V. J. G., J. Uittenbogaard, and P. Pellen. 1996. Wageningen evaluating programmes for analytical laboratories (WEPAL) organizations and purpose. *Commun. Soil Sci. Plant Anal.* 27:421–431.

Hesse, P. R. 1971. *A Textbook of Soil Chemical Analysis.* New York: Chemical Publishing Company.

Ihnat, M. 1993. Reference materials for data quality, In: *Soil Sampling and Methods of Analysis,* M. R. Carter (ed.), Boca Raton, FL: Lewis Publishing, 247–262.

Ingle, J. D. and S.R. Crouch. 1988. *Spectrochemical Analysis.* Englewood Cliffs, NJ: Prentice Hall.

Isaac, R. A. and J. D. Kerber. 1971. Atomic and flame photometry: Techniques and uses in soil, plant, and water analysis, In: *Instrumental Methods for Analysis of Soils and Plant Tissue,* L. M. Walsh (ed.). Madison, WI: Soil Science Society of America, 17–38.

Isaac, R. A. and J. B. Jones, Jr. 1971. AutoAnalyzer systems for the analysis of soil and plant tissue extracts, In: *Technicon International Congress,* Tarrytown, NY: 57–64.

Jackson, M. L. 1958. *Soil Chemical Analysis.* Englewood Cliffs, NJ: Prentice-Hall.

Jones, Jr., J. B. 1977. Elemental analysis of soil extracts and plant tissue ash by plasma emission spectroscopy. *Commun. Soil Sci. Plant Anal.* 8:345–365.

Karmarkar, S. V. 1998. Enhanced ion chromatography with sequential flow injection analysis: Determination of common anions and nitrite. *Amer. Environ. Lab.* 10:1, 6–7.

Lajunen, L. H. J. 1992. *Spectrochemical Analysis by Atomic Absorption and Emission.* Cambridge, England: The Royal Society of Chemistry.

Knopp, J. F. and D. McKee. 1983. *Methods for Chemical Analysis of Water and Wastes.* EPA-600/4-79-020. Cincinnati, OH: Environmental Protection Agency.

Mavrodineanu, R. (ed.). 1970. *Analytical Flame Spectroscopy.* New York: Springer-Verlag.

McLain, D. B. (ed.). 1982. *Handbook for AOAC Members,* 5th ed. Arlington, VA: Association of Official Analytical Chemists.

Metcalfe, E. 1987. *Atomic Absorption and Emission Spectroscopy.* New York: John Wiley & Sons.

Miller, R. O. and J. Kotuby-Amacher, and N. B. Dellevalle. 1996. A proficiency testing program for the agricultural laboratory industry: Results for the 1994 program. *Commun. Soil Sci. Plant Anal.* 27:451–461.

Mills, H. A. 1980. Nitrogen specific-ion electrodes for soil, plant, and water analysis. *J. Amer. Assoc. Anal. Chem.* 63:797–801.
Mitchell, R. L. 1964. *The Spectrographic analysis of Soil, Plants, and Related Materials.* Technical Communication No. 44A of the Commonwealth Bureau of Soils, Harpenden. Bucks, England: Commonwealth Agricultural Bureaux, Farnham Royal.
Montasar, A. and D. W. Golightly (eds). 1987. *Inductively Coupled Plasma in Analytical atomic Spectrometry.* New York: VCH Publishers, Inc.
Morrison, G. H. and T. H. Risby. 1979. Elemental trace element analysis of biological materials. *CRC Critical Rev. Anal. Chem.* 8:287–320.
Munter, R. C. and R. A. Grande. 1981. Plant tissue and soil extract analysis by ICP-atomic emission spectrometry, In: *Developments in Atomic Plasma Spectrochemical Analysis,* R. M. Barmes (ed.), London: Heyden and Son, Ltd., 653–672.
Paul, Steven. 1998. UV-VIS instruments continue to envolve. *Today's Chem. at Work* 7(9):21–25.
Petterelli, M.N. 1995. Axial ICP for trace metal analysis. *Environ. Test Anal.* 13:50–54.
Pickett, E. E. and S. R. Koirtyohann. 1969. Emission flame photometer: A new look at an old method. *Anal. Chem.* 41:28A–30A.
Piper, C. S. 1942. *Soil and Plant Analysis.* Adelaide, Australia: The University of Adelaide.
Quevauviller, P. 1996. Certified reference materials for the quality control of total and extractable trace element determinations in soils and sludges. *Commun. Soil Sci. Plant Anal.* 27:403–418.
Ranger, C. 1981. Flow injection analysis — principles, techniques, applications, design. *Anal. Chem.* 53:20A–26A.
Ravindranath, R. *Principles and Practice of Chromatography.* New York: John Wiley & Sons
Rodriquez, J. B., J. R. Self, and P. N. Soltanpour. 1994. Optical conditions for phosphorus analysis by the ascorbic acid-molybdenum blue method. *Soil Sci. Soc. Amer. J.* 58:866–870.
Rump, H. H. and H. Krist. 1988. *Laboratory Manual for the Examination of Water, Waste Water, and Soil.* New York: VCH Publishers.
Ruzicka, J. and E. H. Hansen. 1988. *Flow Injection Analysis.* Second Edition. New York: John Wiley & Sons.
Scoog, D. A., D. M. West, and F.J. Holler. 1988. *Fundamentals of Analytical Chemistry,* 5th ed. New York: Saunders.
Scott, R. N., V.A. Fassel, R. N. Kniseley, and D. E. Dixon. 1976. Inductively coupled plasma-optical emission analytical spectroscopy. *Anal. Chem.* 46:75–80.
Sharp, B. L. 1991. Inductively coupled plasma spectrometry, In: *Soil Analysis: Modern Instrumental Techniques,* 2nd ed., K. A. Smith (ed.). New York: Marcel Dekker, Inc. 63–109.
Singh, R., D. K. Bhumbia, and R. F. Keefer. 1995. Recommended soil sulfate-S tests, In: *Recommended Soil Testing Procedures for the Northeastern United States.* Northeast Regional Publications No. 493 (revised). J.T. Sims and A. M. Wolf (eds.). Newark DE: University of Delaware, 46–50.
Skogerboe, R. K., I. T. Urasa, and G. N. Coleman. 1976. Characteristics of a DC plasma as an excitation source for multielement analysis. *Appl. Spectrosc.* 30:500–507.
Smith, K. A. (ed). 1991. *Soil Analysis: Modern Instrumental Techniques,* 2nd ed. New York: Marcel Dekker.
Smith, K. A. and A. Scott. 1991. Continuous-flow and descrite analysis, In: *Soil Analysis: Modern Instrumental Techniques,* 2nd ed., K. A. Smith (ed.). New York: Marcel Dekker, 183–227.
Smoley, C. K. 1992. *Methods for the Determination of Metals in Environmental Samples.* Boca Raton, FL: CRC Press LLC.

Soltanpour, P. N., G. W. Johnson, S. M. Workman, J. B. Jones, Jr., and R. O. Miller. 1996. Inductively coupled plasma emission spectrometry and inductively coupled plasma-mass spectrometry, In: *Methods of Soil Analysis, Part 3, Chemical Methods,* SSSA Book Series No. 5. D. L. Sparks (ed.), Madison, WI: Soil Science Society of America, 91–139.

Soltanpour, P. N., C. W. Johnson, S. M. Workman, J. B. Jones, Jr., and R. O. Miller. 1998. Advances in ICP emission and ICP mass spectrometry. *Adv. Agron.* 64:28–113.

Tabatabai, M. A. and N. T. Basta. 1991. Ion chromatography, In: *Soil analysis: Modern Instrumental Techniques,* 2nd ed., K. A. Smith (ed.). New York: Marcel Dekker, 229–259.

Tabatabai, M. A. and W. T. Frankenberger, Jr. 1996. Liquid chromatography, In: *Methods of Soil Analysis, Part 3, Chemical Methods.* SSSA Book Series No. 5, R. L. Sparks (ed.). Madison, WI. Soil Science Society of America, 225–245.

Talibudeen, O. 1991. Ion-selective electrodes, In: *Soil Analysis: Modern Instrumental Techniques,* 2nd ed., K. A. Smith (ed.). New York: Marcel Dekker, 111–182.

Taylor, J. K. 1987. *Quality Assurance of Chemical Measurements.* Chelsea, MI: Lewis Publishers.

Thompson, M. and J. N. Walsh. 1983. *A Handbook of Inductively Coupled Plasma Spectrometry.* Glasgow: Blackie & Son.

Ure, A. M. 1991. Atomic absorption and flame emission spectrometry, In: *Soil Analysis: Modern Instrumental Techniques,* 2nd ed., K. A. Smith (ed.). New York: Marcel Dekker, 1–62.

van Dijk, D. V. J. G. Houba, and J.P.J. van Dalen. 1996. Aspects of quality assurance within the Wageningen evaluating programmes for analytical laboratories. *Commun. Soil Sci. Plant Anal.* 27:433–439.

van Loon, J. C. 1985. *Selected Methods of Trace Metal Analysis; Biological and Environmental samples.* New York: John Wiley & Sons.

Varma, Asha. 1991. *Handbook of Inductively Coupled Plasma Atomic Emission Spectroscopy.* Boca Raton, FL: CRC Press, Inc.

Walsh, L. M. (Ed.). 1971. *Instrumental Methods for Analysis in Soils and Plant Tissue,* Revised Edition. Madison, WI: Soil Science Society of America.

Watson, M. E. and R. A. Isaac. 1990. Analytical instruments for soil and plant analysis, in: *Soil Testing and Plant Analysis,* 3rd ed., SSSA Book Series No. 3. R. L. Westerman (ed.). Madison, WI: Soil Science Society of America, 691–740.

Wolf, A. M. and R. O. and Miller. 1998. Development of a North American Proficiency Testing Program for soil and plant analysis. *Commun. Soil Sci. Plant Anal.* 28:1685–1690.

Wright, R. J. and T. I. Stuczynski. 1996. Atomic absorption and flame emission spectrometry, In: *Methods of Soil Analysis, Part 3, Chemical Methods,* SSSA Book Series No. 5, D. L. Sparks (Ed.). Madison, WI: Soil Science Society of America, 65–90.

Appendix A

REAGENTS, STANDARDS, pH BUFFERS, ACIDS, INDICATORS, STANDARD ACIDS, BASES, AND BUFFERS CITED IN THE HANDBOOK

REAGENTS

aluminum chloride ($AlCl_3 \cdot 6H_2O$)
ammonium acetate ($NH_4C_2H_3O_2$)
ammonium bicarbonate (NH_4HCO_3)
ammonium chloride (NH_4Cl)
ammonium fluoride (NH_4F)
ammonium hydroxide (NH_4OH)
ammonium molybdate $[(NH_4)_6Mo_7O_{24} \cdot 4H_2O]$
ammonium nitrate (NH_4NO_3)
ammonium dihydrogen phosphate ($NH_4H_2PO_4$)
antimony potassium tartrate $[K(SbO)C_4H_4O_6 \cdot 1/2H_2O]$
ascorbic acid ($C_6H_8O_6$)
azomethine-H

barium chloride ($BaCl_2 \cdot 2H_2O$)
boric acid (H_3BO_3)

calcium acetate $[Ca(C_2H_3O_2)_2]$
calcium carbonate ($CaCO_3$)
calcium chloride ($CaCl_2 \cdot 2H_2O$)
calcium dihydrogen phosphate $[Ca(H_2PO_4)_2]$
calcium sulfate ($CaSO_4 \cdot 5H_2O$)
calcium hydroxide $[Ca(OH)_2]$
citric acid ($C_6H_8O_7$)
copper sulfate ($CuSO_4 \cdot 5H_2O$)

DTPA—diethylenetriaminepentaacetic acid
DTPA—diethylenetriaminepentaacetic acid (penta sodium salt)
    4,5-dihydroxy-2,7-naphthalene-disulfonic acid (disodium salt)
    $[(HO)_2C_{10}H_4(SO_3Na)_2]$
disodium phosphate ($Na_2HPO_4 \cdot 12H_2O$)
disodium salt of ethylenedinitrilo tetraacetic acid
disodium salt of nitrilotriacetic acid

ethyl alcohol, denatured (C_2H_5OH)
EDTA—ethylenediaminetetraacetic acid
(ethylenedinitrillo) tetraacetic acid-tetrasodium salt

ferric nitrate [Fe(NO$_3$)$_3$·9H$_2$O]
ferrous ammonium sulfate [Fe(NH$_4$)$_2$SO$_4$·6H$_2$O]

hydrogen peroxide (H$_2$O$_2$)
hydroxylamine HCl

lanthanum nitrate hexahydrate [La(NO$_3$)$_3$·6H$_2$O]
lanthanum oxide (La$_2$O$_3$)
lithium chloride (LiCl)

manganese oxide (MnO$_2$)
mercury thiocyanate [Hg(SCN)$_2$]

a-naphtylamine (C$_{10}$H$_9$N)
o-naphtyletyhylene-diamine dihydrochloride (C$_{12}$H$_{16}$Cl$_2$N$_2$)
nitrilotriacetic acid-disodium salt

p-nitrophenol (HO·C$_6$H$_4$·NO$_2$)
paranitrophenol
n-phenylanthranilic acid
potassium aluminum sulfate [KAl(SO$_4$)$_2$·12H$_2$O]
potassium antimony tartrate (KSbOC$_4$H$_4$O$_6$·5H$_2$O)
potassium chloride (KCl)
potassium chromate (KCrO$_4$)
potassium dichromate (K$_2$Cr$_2$O$_7$)
potassium dihydrogen phosphate (KH$_2$PO$_4$)
potassium hydroxide (KOH)
potassium nitrate (KNO$_3$)
potassium persulfate (K$_2$S$_2$O$_8$)

silver sulfate (Ag$_2$SO$_4$)
sodium acetate (NaC$_2$H$_3$O$_2$)
sodium bicarbonate (NaHCO$_3$)
sodium borate (Na$_2$B$_4$O$_7$·10H$_2$O)
sodium carbonate (Na$_2$CO$_3$)
sodium chloride (NaCl)
sodium dichromate (Na$_2$Cr$_2$O$_7$)
sodium glycerophosphate [Na$_2$C$_3$H$_5$(OH)$_2$PO$_4$·5 1/2H$_2$O]
sodium hydroxide (NaOH)
sodium nitroprusside dihydrate {Na$_2$-[Fe(CN)$_5$NO]·H$_2$O}
sodium potassium tartrate (C$_2$H$_4$KNaO$_6$)
sodium salicylate (C$_7$H$_5$O$_3$Na)
sodium tetraborate (Na$_2$B$_4$O$_7$·10H$_2$O)
stannous chloride (SnCl$_2$·2H$_2$O)
sulfanilamide (C$_6$H$_8$N$_2$O$_2$S)

triethanolamine (TEA)

zinc sulfate (ZnSO$_4$·7H$_2$O)

Appendix A

FOR PREPARATION OF STANDARDS

ammonium dihydrogen phosphate ($NH_4H_2PO_4$)

boric acid (H_3BO_3)

cadmium (Cd) metal
cadmium nitrate [$Cd(NO_3)_2 \cdot H_2O$]
calcium carbonate ($CaCO_3$)
copper (Cu) metal

humic acid

iron (Fe) wire

lead nitrate [$Pb(NO_3)_2$]

magnesium (Mg) ribbon
magnesium sulfate ($MgSO_4 \cdot 7H_2O$)
manganese oxide (MnO_2)

nickel (Ni) metal
nickel nitrate [$Ni(NO_3)_2 \cdot 6H_2O$]

potassium chloride (KCl)
potassium hydrogen phthalate ($KC_8H_5O_4$)
potassium sulfate (K_2SO_4)

sodium chloride (NaCl)

zinc (Zn) metal

REAGENTS FOR pH BUFFERS

citric acid ($C_6H_8O_7$)

disodium phosphate ($Na_2HPO_4 \cdot 12H_2O$)

ACIDS

acetic acid ($HC_2H_3O_2$)

hydrochloric acid (HCl)

nitric acid (HNO_3)

perchloric acid ($HClO_4$)
phosphoric acid (H_3PO_4)

sulfuric acid (H_2SO_4)

INDICATORS

bromocresol green

methyl red
methylene blue

phenolphthalein

n-phenylanthranilic acid

STANDARD ACIDS, BASES, AND BUFFERS

Normality of Concentrated Acids and Bases

Acids	Specific Gravity	Percent By Weight	Normality
Concentrated H_2SO_4	1.84	95.5–96.5	36.02
Concentrated HCl	1.19	37.6	12.27
Concentrated HNO_3	1.423	70.7	15.96
Glacial Acetic Acid ($HC_2H_3O_2$)	1.049	100	17.47
Concentrated H_3PO_4	1.689	86.3	44.62
Bases			
Concentrated NH_4OH	0.899	28.8	7.39
Ammonia	28.0	14.5	69.0
Sodium Hydroxide (NaOH)	50.5	19.4	51.5
Potassium Hydroxide (KOH)	45.0	11.7	85.5

Normal Solution Preparation

(To make various normal solutions, dilute aliquots of concentrated acid or base to 1,000 mL with deionized water)

Acids	mL of concentrated acid to make 1,000 mL				
	0.1N	1.0N	2.0N	5.0N	10.0N
Conc. H_2SO_4	2.78	27.76	55.52	138.81	277.626
Conc. HCl	8.15	81.50	163.00	407.50	815.00
Conc. HNO_3	6.26	62.66	125.31	313.28	626.56
Conc. NH_4OH	13.53	135.32	270.63	676.59	—
Glacial $HC_2H_3O_2$	5.72	57.24	114.48	286.2	572.4
Conc. H_3PO_4	2.24	22.41	44.82	112.06	224.11

BUFFER SOLUTIONS

Stock Solutions (use reagent grade chemicals and pure water)

- a. ***0.2 M Acid Potassium Phthalate*** ($KHC_8H_4O_4$): Dry $KHC_8H_4O_4$ to constant weight at 110° to 115°C. Dissolve 40.836 g in water and dilute to 1 liter with water.
- b. ***0.2 M Monopotassium Phosphate*** (KH_2PO_4): Dry KH_2PO_4 to constant weight at 110° to 115°C. Dissolve 27.232 g in water and dilute to 1 liter with water.

Appendix A

   c. **0.2 M Boric Acid Potassium Chloride** H_3BO_3-(KCl): Dry H_3BO_3 to constant weight in desiccator over anhydrous calcium chloride ($CaCl_2$). Dry KCl two days in an oven at 115° to 120°C. Dissolve 12.405 g H_3BO_3 and 14.912 g KCl in water and dilute to 1 liter with water.
   d. **0.2 M Sodium Hydroxide** (NaOH): To 1 part NaOH add 1 part water. Mix to dissolve and let stand until clear (about 10 days). Dilute 10.8 mL of this solution to 10 liters with water. Titrate against weighed amount of acid potassium phthalate ($KHC_8H_4O_4$). 0.04084 g $KHC_8H_4O_4$ = 1 mL 0.2M NaOH. It is preferable to use factor with solution rather than try to adjust to exactly 0.2M.

Preparation of Buffer Solutions

(Prepare standard buffer solutions from designated amounts of stock solutions and dilute to 200 mL with water)

Phthalate-NaOH Mixtures

pH of Mixture	0.2M $KHC_8H_4O_4$ mL	0.2M NaOH mL
5.0	50	23.65
5.2	50	29.75
5.4	50	35.25
5.6	50	39.70
5.8	50	43.10
6.0	50	45.40
6.2	50	47.00

KH_2PO_4-NaOH Mixture

pH of Mixture	0.2M KH_2PO_4 mL	0.2M NaOH mL
5.8	50	3.66
6.0	50	5.64
6.2	50	8.55
6.4	50	12.60
6.6	50	17.74
6.8	50	23.60
7.0	50	29.54
7.2	50	34.90
7.4	50	39.34
7.6	50	42.74
7.8	50	45.17
8.0	50	46.85

Appendix B

STANDARDS AND STANDARD PREPARATION

Preparation

Carefully prepared calibration standards are required to assay soil extractants no matter what instrumental procedure is used. The two essential requirements for these standards are that (i) their composition is as closely matched to the unknowns as possible, and (ii) their elemental concentration range bracket is that expected to be found in the unknowns.

There are two types of standards, Primary Standards and Working Standards. A Primary Standard normally is a single element standard at a specific concentration, such as 100, 1,000, or 10,000 mg L^{-1}. Primary Standards may be made by the analyst (see table in this section) or purchased from a commercial company (see list at the end of this Appendix). Working Standards on the other hand, are prepared from Primary Standards by the analyst. A Primary Standard may be made in the matrix (such as the extraction reagent) for its expected use, while the Working Standards must always be prepared in the same matrix as the unknowns.

Single element-containing Working Standards are relatively easy to prepare. An example would be the preparation of a series of calibration standards for the spectrophotometric (colorimetric) determination of phosphorus (P) in a soil extract by the phosphomolybdate blue procedure. The working concentration range must be determined and standards prepared to cover that range. Normally, the useful concentration range for most spectrophotometric procedures is one decade (0 to 10 mg L^{-1}), or possibly two (0 to 100 mg L^{-1}). In addition, it will be necessary to determine either by preliminary testing or from past experience what elemental concentration range will be expected in the unknowns. If the concentration range of unknowns is not within the working range of the spectrophotometric procedure chosen, then the unknowns must be diluted to bring them into that working range.

The greater the number of Working Standards, the more precise the calibration curve will be. However, there are practical issues that must be considered. Five Working Standards, including the zero standard (that is, a standard without the element of interest), are normally a sufficient number. For example, a series of five Working Standards for the P determination in a soil extractant ranging in concentration from 0 to 10.0 mg L^{-1} would normally cover the working range for the phosphomolybdate blue procedure (see Chapter 6). Depending on the aliquot of Working Standard used, the size of the cuvette (that is, the length of the light path through the cuvet), and the characteristics of the spectrophotometer, the working range may be decreased or increased. Hopefully, the P concentration of all unknowns will fall below the highest P concentration Working Standard. If not, then those high P containing unknowns must be diluted and their P concentrations determined using another set of Working Standards prepared so as to match the diluted unknowns.

The same general procedure would be followed for multielement standards, standards required when the same analytical instrument can assay an unknown for more than one element. For example, standards for calibrating flame emission for potassium (K) and sodium (Na) and for atomic absorption spectrophotometry for calcium (Ca), magnesium (Mg), copper (Cu), iron (Fe), manganese (Mn), and zinc (Zn) could contain more than one element, thereby simplifying their preparation and use. Normally, the working concentration range would be one to two decades which must be determined by testing or it may already be known. Then, the concentration range in the unknowns must be determined. Dilution of the unknowns may be necessary so that their elemental content will fall within the working range of the analytical procedure. When diluting samples, it is important that the obtained diluted sample be identical in matrix composition to the standards.

For multielement Working Standards, it is not wise to have every element in the series at the same increasing concentration so that in one standard all the elements are at the lowest and in another are all at the highest concentrations. It may also be necessary to alternate the concentrations of various elements in the Working Standards so that there is a minimum of possible interaction among the elements that might result in precipitation or some undesirable complex formation.

The first essential requirement is that the composition of Working Standards closely matches the unknowns whose elemental contents will be determined by comparison to the Working Standards. This means that for soil extractants, the Working Standards should be made in the extraction reagent. If fairly large aliquots of the Primary Standards are needed, they should also be made in the extraction reagent. This requirement is to minimize what is known as the *Matrix Effect*.

MATRIX EFFECTS

The *Matrix Effect* is an influence that the chemical and physical characteristics of a standard or sample unknown can have on its analysis. For example, the total ion content and type, pH, presence of organics, color, viscosity, etc., can affect the way a standard or unknown will react when assayed.

MATRIX MODIFIERS

In some types of elemental determinations, most frequently assays by both flame and graphite furnace atomic absorption spectrophotometry (AAS and GF-AAS, respectively), a *Matrix Modifier* is needed to ensure that the elemental species being detected or utilized is present. For example, in the determination of Ca by AAS, the presence of other companion ions, such as aluminum, phosphate and sulfate, and as well as organic substances in the analyte, can combine with the element, reducing the concentration of the species (ground-state Ca atoms) being measured in the flame or gaseous material, resulting in an erroneous (normally low) assay result. To prevent such combining from taking place, the addition of a Matrix Modifier, in this case lanthanum (La), and sometimes strontium (Sr), will prevent this combining from taking place. For more details on this subject, refer to the chapters by Watson and Isaac (1990) and Wright and Stuczynski (1996).

BLANKS

A *Blank* is obtained by carrying forward the analysis but without interaction with a sample. For example, carrying forward an extraction without interaction with a soil sample would produce a Blank. A Blank serves a very useful purpose by determining, if in the preparation of unknowns, an element(s) to be determined in the assay is being added systematically to the unknowns. A standard of the matrix only (void of analyte elements) is not a Blank and should not be so used. However, an assay of reagents used in the assay should be carefully checked for their freedom of the analyte(s) being determined. Blanks should be prepared when beginning a series of sample preparations and periodically thereafter, particularly when a change in the procedure occurs. High Blank values indicate a significant source of contamination and may be sufficient to invalidate a determination. Some have suggested that Working Standards might be processed in the same manner as the unknown samples so as to equally contaminate both standards and unknowns. Such a procedure is dangerous since elimination of the source of contamination is the proper solution.

PREPARATION OF WORKING STANDARDS FOR THE INDUCTIVELY COUPLED PLASMA (ICP) EMISSION SPECTROMETER

Properly prepared standards are required in order to perform analyses on an inductively coupled plasma (ICP) emission spectrometer. Accuracy is best determined by the use of Primary Standards, standards with known composition verified by certification, such as the Standard Reference Materials (SRMs) issued by the U.S. National Institute of Standards and Technology (NIST, Gaithersburg, MD 20899). Unfortunately, SRMs are not available for every sample type or all elements. Today numerous Primary and Working Standards can be purchased from commercial firms that specialize in preparing and marketing standards for wide ranges of use, including quality assurance applications. With these sources of reliable standards, analysts are not without means of verifying accuracy. For daily use in calibration, the analyst would normally prepare other types of standards, frequently referred to as Working Standards. In most instances, these standards are prepared by the analyst who may or may not be able to verify their element content accurately. In cases where verification of accuracy is not possible, some commercial standard providers can prepare Working Standards suited for use with the types of samples being assayed.

Care must be taken to ensure that the Working Standards prepared to calibrate an ICP, or any other analytical instrument, are as similar in composition and elemental concentration range as those being assayed as unknowns. For example, Working Standards for the assay of a soil extractant must be prepared in the Extracting Reagent with the elemental concentration range coverage being that expected to occur in the unknowns.

The analyst can prepare his own Primary Standard or use commercially available 100, 1,000, or 10,000 mg L^{-1} elemental standards for the preparation of Working Standards. It is good laboratory practice to analyze both prepared or purchased 100, 1,000, or 10,000 mg L^{-1} Primary Standards before using them in order to verify their elemental content as well as to determine the presence of other than the designated element(s). In order to minimize possible incompatibility, it is best to prepare several Working Standards, grouping certain elements together while keeping others separate. The objective is to prepare Working Standards that have a reasonably good shelf life as well as conforming to the requirements in terms of their composition and elemental concentration.

NEED FOR AN INTERNAL STANDARD(S)

The requirement and use of an Internal Standard(s) when assays are done by means of sample aspiration (flame, atomic absorption spectrophotometry, and ICP spectrometry) are dependent on the efficiency of sample transfer. Factors, such a liquid viscosity that would alter the flow rate through a nebulizer, or the presence of substances that would alter the shape or temperature profile of a flame or plasma (such as organic substances or one dominant ion), influence emission intensity.

One solution to this problem is the use of an Internal Standard. The concept and use of Internal Standards are not new. The Internal Standard has been a common feature on most precision flame photometers. The assumption is that whatever occurs in the aspiration rate, or change of flame or plasma configuration on emission intensity equally happens to an added Internal Standard. However, the use of a peristaltic pump to introduce the liquid sample into the nebulizer provides some control of the analyte flow rate.

For the assay of water and most soil extracts by ICP emission with peristaltic pumping of the analyte into the nebulizer, an Internal Standard is probably not needed. It would be well, however, to compare analysis precision with and without the use of an Internal Standard. For best results, the Internal Standard should be similar in terms of its excitation potential to the element(s) being assayed. For example, the optimum internal standard for the determination of K and Na, in particular, as well as Ca and Mg, is lithium (Li). For the heavy metals, such as Cu, Fe, Mn, Zn, etc., an element (not detectable in the analyte) that has similar chemical and emission characteristics should be chosen. For the assay of a complex matrix, more than one Internal Standard may be required to cover all the various elements being determined. The Internal Standard should be added to all standards and unknowns in sufficient concentration to be equal to or slightly greater in concentration than the concentration of the element to be determined in the unknown. For example, if Li is the Internal Standard for determining K found at 1,000 mg L^{-1} in the unknowns, then the Li concentration should be at least 1,000 mg L^{-1}.

Experience has shown that an Internal Standard may significantly improve the precision of an assay for substances having wide ranges in elemental composition.

Appendix B

Primary Standard Solution Preparation (source: Ward, 1978)[†].

Element	Reagent	Weight, g	Solvent
Aluminum (Al)	Al	1.0000	6 M HCl
	$AlCl_3 \cdot 6H_2O$	8.9481	1 M HCl
Arsenic (As)	As	1.0000	4 M HCl
	As_2O_3	1.3203	4 M HCl
Barium (Ba)	$BaCl_2$[††]	1.1516	Water
	$BaCO_3$[††]	1.4369	0.05 M HNO_3
	$Ba(NO_3)_2$	1.9029	Water
Boron (B)	H_3BO_3	5.7191	Water
Cadmium (Cd)	Cd	1.0000	4 M HNO_3
	CdO	1.1423	4 M HNO_3
Calcium (Ca)	$CaCO_3$	2.4972	0.5 M HNO_3
	$Ca(NO_3)_2 \cdot 4H_2O$[††]	5.8920	Water
Chromium (Cr)	Cr	1.00011	4 M HCl
	$CrCl_3 \cdot 6H_2O$	5.1244	Water
Cobalt (Co)	Co	1.0000	4 M HCl
	$CoCl_2 \cdot 6H_2O$	4.0373	Water
Copper (Cu)	Cu	1.0000	4 M HNO_3
	CuO	1.2518	4 M HNO_3
Iron (Fe)	Fe	1.0000	4 M HCl
	Fe_2O_3	1.4297	4 M HCl
Lead (Pb)	Pb	1.0000	4 M HNO_3
	PbO	1.0772	4 M HNO_3
	$Pb(NO_3)_2$	2.6758	Water
Magnesium (Mg)	Mg	1.6581	0.5 M HCl
	$MgCl_2 \cdot 6H_2O$[††]	8.3621	Water
Manganese (Mn)	Mn	1.0000	4 M HNO_3
	MnO_2	1.5825	4 M HNO_3
Molybdenum (Mo)	Mo	1.0000	*Aqua regia*
	MoO_3	1.5003	*Aqua regia*
Nickel (Ni)	Ni	1.0000	4 M HCl
	NiO	1.2725	4 M HCl
	$NiCl_2 \cdot 6H_2O$	4.0489	Water
Phosphorus (P)	NaH_2PO_4	3.8735	Water
	$NaNH_4H_2PO_4$	3.7137	Water
Potassium (K)	KCl	1.9067	Water
	K_2CO_3	1.7673	1 M HCl
Selenium (Se)	SeO_2	1.4053	Water
Sodium (Na)	NaCl	2.5421	Water
	Na_2CO_3	2.3051	1 M HCl
Zinc (Zn)	Zn	1.0000	4 M HNO_3
	ZnO	1.2448	4 M HNO_3
	$Zn(NO)_2 \cdot 6H_2O$	4.5506	Water

[†]Use 100 to 150 mL of solvent to dissolve and bring to 1,000 mL volume to give a concentration of 1,000 ppm of element.

[††]Not Specpure materials.

REFERENCES

Ward, A. F. 1978. Stock standards preparation. *Jarrell-Ash Plasma Newsletter* 1(2):14–15.

Watson, M. E. and R. A. Isaac. 1990. Analytical instruments for soil and plant analysis, In: *Soil Testing and Plant Analysis,* SSSA Book Series No. 3, R. L. Westerman (ed.). Madison, WI: Soil Science Society of America, 691–740.

Wright, R. J. and T. I. Stuczynski. 1996. Atomic absorption and flame emission spectrometry, In: *Methods of Soil Analysis, Part 3 Chemical Analysis,* SSSA Book Series No. 5, R. L. Sparks (ed.). Madison, WI: Soil Science Society of America, 65–90.

SOURCES OF STANDARDS

AccuStandard, Inc., 125 Market Street, New Haven, CT 06513 (800-442-5290)

Hawk Creek Laboratory, R.D. 1 Box 686, Simpson Road, Glen Rock, PA 17327 (800-637-2436)

National Institute of Standards and Technology (NIST), Standard Reference Materials Program, Room 204, Building 202, Gaithersburg, MD 20899-0001 (301-975-6776)

NSI Solutions, Inc., 2 Triangle Drive, Research Triangle Park, NC 27709 (800-234-7837)

Spex Chemical Division, 203 Harcross Avenue., Metuchen, NJ 08840 (908-549-71454)

Appendix C

CONVERSION FACTORS

Common Prefixes

Factor	Prefix	Symbol
1,000,000	mega	M
1,000	kilo	k
1/100	centi	c
11/000,	milli	m
1/1,000,000	micro	µ

Metric Conversion Factors (approximate)

	When you know	Multiply by	To find	Symbol
Length	inches	2.54	centimeters	cm
	feet	30	centimeters	cm
	yards	0.9	meters	m
	miles	1.6	kilometers	km
Area	square inches	6.5	square centimeters	sq cm
	square feet	0.09	square meters	sq m
	square yards	0.8	square meters	sq m
	square miles	2.6	square kilometers	sq km
	acres	0.4	hectares	ha
Weight	ounces	28	grams	g
	pounds	0.45	kilograms	kg
	short tons (2,000 pounds)	0.9	metric tons	t
Volume	teaspoons	5	milliliters	mL
	tablespoons	15	milliliters	mL
	cubic inches	16	milliliters	mL
	fluid ounces	30	milliliters	mL
	cups	0.24	liters	L
	pints	0.47	liters	L
	quarts	0.95	liters	L
	gallons	3.8	liters	L
	cubic feet	0.03	cubic meters	cu m
	cubic yards	0.76	cubic meters	cu m
Pressure	inches of mercury	3.4	kilopascals	kPa
	pounds/square inch	6.9	kilopascals	kPa
Temperature (exact)	degrees Fahrenheit (after subtracting 32)	5/9	degrees Celsius	°C

Useful Information and Conversion Factors

Name	Symbol	Approximate Size or Equivalent
Length		
meter	m	39.5 inches
kilometer	km	0.6 mile
centimeter	cm	width of a paper clip
millimeter	mm	thickness of a paper clip
Area		
hectare	ha	2.5 acres
Weight		
gram	g	weight of a paper clip
kilogram	kg	2.2 pounds
metric ton	t	long ton (2,240 pounds)
Volume		
liter	L	one quart and 2 ounces
milliliter	mL	1/5 teaspoon
Pressure		
kilopascal	kPa	atmospheric pressure is about 100 kPa
Temperature		
Celsius	C	5/9 F after subtracting 32 from °F
freezing	0°C	32°F
boiling	100°C	212°F
body temp.	37°C	98.6°F
room temp.	20–25°C	68–77°F
Electricity		
kilowatt	kW	
kilowatt-hour	kWh	
megawatt	MW	
Miscellaneous		
hertz	Hz	one cycle per second

Yield or Rate

Ounces per acre (oz A^{-1}) × 0.07 = kilograms per hectare (kg ha^{-1})
Tons per acre (ton A^{-1}) × 2240 = kilograms per hectare (kg ha^{-1})
Tons per acre (ton A^{-1}) × 2.24 = metric tons per hectare (kg ha^{-1})
Pounds per acre (lb A^{-1}) × 1.12 = kilograms per hectare (kg ha^{-1})
Pounds per cubic foot (lbs ft^{-3}) × 16.23 = kilograms per cubic meter (kg m^{-3})
Pounds per gallon (lbs gal^{-1}) × 0.12 = kilograms per liter (kg L^{-1})
Pounds per ton (lbs ton^{-1}) × 0.50 = kilograms per metric ton (kg MT^{-1})
Gallons per acre (gal A^{-1}) × 9.42 = liters per hectare (L ha^{-1})
Gallons per ton (gal ton^{-1}) × 4.16 = liters per metric ton (L MT^{-1})
Pounds per 100 square foot (lbs ft^{-3}) × 2 = pounds per 100 gallons water
  (assumes that 100 gallons will saturate 200 square foot of soil)
Pounds per acre (lbs A^{-1})/43.56 = lbs per 1000 square foot (lbs ft^{-2})

Appendix C

Volumes and Liquids

1 teaspoon = 1/3 tablespoon = 1/16 ounce
1 tablespoon = 3 teaspoons = 1/2 ounce
1 fluid ounces (oz) = 2 tablespoons = 6 teaspoons
1 pint per 100 gallons = 1 teaspoon per gallon
1 quart per 100 gallons = 2 tablespoons per gallon
3 teaspoons = 1 tablespoon (tsp) = 14.8 milliliters (mL)
2 tablespoons (tsp) = 1 fluid ounces = 29.6 milliliters (mL)
8 fluid ounces (oz) = 16 tablespoons (tsp) = 1 cup = 236.6 milliliters (mL)
2 cups = 32 tablespoons (tsp) = 1 pint = 473.1 milliliters (mL)
2 pints = 64 tablespoons (tsp) = 1 quart (qt) = 946.2 milliliters (mL)
1 liter (L) = 1,000 milliliters (mL) = 1,000 cubic centimeters (cc) =
    0.264 gallons (gal) = 33.81 ounces (oz)
4 quarts (qt) = 256 tablespoons (tsp) = 1 gallon (gal) = 3785 milliliters (mL)
1 gallon (gal) = 128 ounces (oz) = 3.785 liters (L)

Elemental Conversions

$P_2O_5 \times 0.437$ = Elemental P   Elemental $P \times 2.29 = P_2O_5$
$K_2O \times 0.826$ = Elemental K   Elemental $K \times 1.21 = K_2O$
$CaO \times 0.71$ = Elemental Ca   Elemental $Ca \times 1.40 = CaO$
$MgO \times 0.60$ = Elemental Mg   Elemental $Mg \times 1.67 = MgO$
$CaCO_3 \times 0.40$ = Elemental Ca

Weight/Mass

1 ounce (oz) = 28.35 grams (g)
16 ounces (oz) = 1 pound (lb) = 453.6 grams (g)
1 kilogram (kg) = 1,000 grams (g) = 2.205 pounds (lb)
1 gallon water = 8.34 pounds (lbs) = 3.8 kilograms (kg)
1 cubic foot of water (cu ft) = 62.4 pounds (lbs) = 28.3 kilograms (kg)
1 kilogram of water (kg) = 33.81 ounces (oz)
1 ton (t) = 2,000 pounds (lbs) = 907 kilograms (kg)
1 metric ton (MT) = 1,000 kilograms (kg) = 2,205 pounds (lbs)

Volume Equivalents

1 gallon in 100 gallons = 1 1/4 ounces (oz) in 1 gallon (gal)
1 quart in 100 gallons = 5/16 oz in 1 gallon (gal)
1 pint in 100 gallon = 3/16 ounces (oz) in 1 gallon (gal)
8 ounces (oz) in 100 gallons = 1/2 teaspoon in 1 gallon (gal)
4 ounces (oz) in 100 gallons = 1/4 teaspoon in 1 gallon (gal)

Temperature

°C	°F	°C	°F
10	50	80	176
20	68	90	194
30	86	100	212
40	104	200	392
50	122	300	573
60	140	400	752
70	158	500	932

Degrees F = (Degrees C + 17.78) × 1.8
Degrees C = (Degrees F − 32) × 0.556

Appendix D

REFERENCE TEXTS

SOIL ANALYSIS

Anon. 1980. *Soil and Plant Testing and Analysis,* FAO Soil Bulletin 38/1. Rome: Food & Agriculture Organization of the United Nations.

Anon. 1989. *Standard Methods for the Examination of Water and Waste Water,* 17th ed. Washington, DC: American Public Health Association.

Brown, J. R. (ed.). 1987. *Soil Testing: Sampling, Correlation, Calibration, and Interpretation,* Special Publication 21. Madison, WI: Soil Science Society of America.

Brown, J. R. (ed.). 1998. *Recommended Chemical Soil Test Procedures for the North Central Region,* North Central Regional Publication 221 (revised). Columbia, MO: Missouri Agricultural Experiment Station SB 1001, University of Missouri.

Carter, M. R. (ed.). 1993. *Soil Sampling and Methods of Analysis.* Boca Raton, FL: CRC Press.

Chapman, H. D. and P. F. Pratt (eds.). 1982. *Methods of Analysis of Soil, Plants, and Waters,* Publication 4034, Berkeley, CA: Division of Agricultural Sciences, University of California.

Dahnke, W. C. (ed.). 1988. *Recommended Chemical Soil Test Procedures for the North Central Region,* North Central Regional Publication 221 (revised), North Dakota Agricultural Experiment Station Bulletin No. 499 (revised), Fargo, ND: North Dakota State University.

Davis, J. and F. Freitas. 1970. *Physical and Chemical Methods of Soil and Water Analysis,* Soils Bulletin 10. Rome: Food & Agriculture Organization of the United Nations.

Donahue, S. J. (ed.). 1983. *Reference Soil Test Methods for the Southern Region of the United States,* Southern Cooperative Series Bulletin 289. Blacksburg, VA: Virginia Agricultural Experimentation Station.

Donohue, S. J. (ed.). 1992. *Reference Soil and Media Diagnostic Procedures for the Southern Region of the United States,* Southern Cooperative Series Bulletin 374. Blacksburg, VA: Virginia Agricultural Experiment Station.

Gavlak, R. G., D. A. Horneck, and R. O. Miller (eds.). 1994. *Plant, Soil, and Water Reference Methods for the Western Region,* Western Regional Extension Publication 125. Fairbanks, AK: University of Alaska Extension Service.

Hardy, G. W. (ed.). 1981. *Soil Testing and Plant Analysis, Soil Testing, Part I,* Special Publication 2. Madison, WI: Soil Science Society of America.

Hesse, R. 1971. *A Textbook of Soil Chemical Analysis.* New York: Chemical Publishing Company, New York, NY.

Jackson, M. L. 1958. *Soil Chemical Analysis.* Englewood Cliffs, NJ: Prentice-Hall.

Johnson, C. M. and A. Ulrich. 1959. *Analytical Methods for Use in Plant Analysis,* Bulletin 766. Berkeley: University of California, Agricultural Experiment Station.

Kalra, Y. P. and D. G. Maynard. 1991. *Methods Manual for Forest Soil and Plant Analysis,* Northwest Region, Information Report NOR-X-319. Edmonton: Forestry Canada.

Klute, A. (ed.). 1996 *Methods of Soil Analysis, Physical, and Mineralogical Properties,* Part 1, 2nd ed., Monograph 9. Madison, WI: American Society of Agronomy.

Loveday J. (ed.). 1974. *Methods for Analysis of Irrigated Soils,* Technician Communication 54. Commonwealth Bureau of Soils, Farmham Royal, Bucks, England.

Page, A. L., R. H. Miller, and D. R. Keeney (eds.). 1982. *Methods of Soil Analysis, Part 2, Chemical and Microbiological Properties,* 2nd ed.. Madison, WI: Soil Science Society of America.

Peverill, K. I., L. A. Sparrow, and D. J. Reuter (eds.). 1999. *Soil Analysis: An Interpretation Manual.* Collingwood, Australia: CSIRO Publishing.

Piper, C. S. 1942. *Soil and Plant Analysis.* Adelaide, Australia: The University of Adelaide.

Radojevic, M. and V. N. Bashkin. 1999. *Practical Environmental Analysis.* Cambridge: The Royal Society of Chemistry.

Sims, J. T. and A. M. Wolf (eds.). 1991. *Recommended Soil Testing Procedures for the Northeastern United States.* Northeast Regional Publication 493. Newark, DE: Agricultural Experiment Station, University of Delaware.

Sparks, D. L. (ed.). 1996. *Methods of Soil Analysis, Part 3, Chemical Method.* Madison, WI: Soil Science Society of America.

Walsh, L. M. and J. D. Beaton (eds.). 1972. *Soil Testing and Plant Analysis.* Revised Edition. Madison, WI: Soil Science Society of America.

Westerman, R. L. (ed.). 1990. *Soil Testing and Plant Analysis.* 3rd ed. Madison, WI: Soil Science Society of America.

GENERAL RELATED TEXTS

Adriano, D. C. 1986. *Trace Elements in the Terrestrial Environment.* New York: Springer-Verlag.

Anon. 1996. *Soil Fertility Manual.* Norcross, GA: Potash & Phosphate Institute.

Barber, S. A. 1995. *Soil Nutrient Bioavailability: A Mechanistic Approach.* 2nd ed. Chichester: John Wiley & Sons.

Bennett, W. F. 1993. *Nutrient Deficiencies & Toxicities in Crop Plants.* St. Paul, MN: The American Phytopathological Society.

Bergman, W. 1992. *Nutritional Disorders in Plants: Development, Visual and Analytical Diagnosis.* Jena, Germany: Gustav Pischer Verlag.

Beverly, R. B. 1991. *A Practical Guide to the Diagnosis and Recommendation Integrated System (DRIS).* Athens, GA: Micro-Macro Publishing.

Black, C. A. 1993. *Soil/Fertility Evaluation and Control.* Boca Raton, FL: Lewis Publishers.

Bould, C., E. J. Hewitt, and P. Needham. 1984. *Diagnosis of Mineral Disorders in Plants: Principles.* Volume 1. New York: Chemical Publishing Co.

Chapman, H. D. 1966. *Diagnostic Criteria for Plants and Soils.* Riverside, CA: Division of Agriculture, University of California.

Childers, N. F. (ed.). 1968. *Fruit Nutrition: Temperate to Tropical.* New Brunswick, NJ: Horticultural Publications, Rutgers—The State University.

Cottenie, A. 1980. *Soil and Plant Testing as a Basis of Fertilizer Recommendations.* Soils Bulletin 38/2. Rome: Food and Agriculture Organization of the United Nations.

Davidescu, D. and V. Davidescu. 1972. *Evaluation of Fertility by Plant and Soil Analysis.* Kent, England: Abacus Press.

Fairbridge, R. W. and C. W. Finkl, Jr. (eds.). 1979. *The Encyclopedia of Soil Science, Physics, Chemistry, Biology, Fertility, and Technology,* Part 1. Stroudsburg, PA: Dowden, Hutchinson & Ross.

Halliday, D. J. and M. E. Trenkel (eds.). 1992. *IFA World Fertilizer Use Manual.* Paris: International Fertilizer Industry Association.

Jones, Jr., J. B., B. Wolf, and H. A. Mills. 1991. *Plant Analysis Handbook: A Practical Sampling, Preparation, Analysis, and Interpretation Guide.* Athens, GA: Micro-Macro Publishing.

Jones, Jr., J. B. 1998. *Plant Nutrition Manual.* Boca Raton, FL: CRC Press.
Kabata-Pendias, Alina and Henryk Pendias. 1995. *Trace Elements in Soils and Plants.* Revised Edition. Boca Raton, FL: CRC Press, Inc.,
Kitchens, H. B. (ed.). 1948. *Diagnostic Techniques for Soils and Crops.* Washington, DC: The American Potash Institute.
Ludwick, A. E. (ed.). 1998. *Western Fertilizer Handbook, Second Horticultural Edition.* Danville, IL: Interstate Publishers.
Maynard, D. N. and G. J. Hochmuth. 1997. *Knott's Handbook for Vegetable Growers.* 4th ed. New York: John Wiley & Sons.
Mortvedt, J. J. (ed.). 1991. *Micronutrients in Agriculture.* Second Edition. Madision, WI: Soil Science Society of America.
Mortvedt, J. J., P. M. Giordano, and W. L. Lindsay (eds.). 1972. *Micronutrients in Agriculture.* Madison, WI: Soil Science Society of America.
Mills, H. A. and J. B. Jones, Jr. 1997. *Plant Analysis Handbook II.* Athens, GA: Micro-Macro Publishing.
Nickolas, D. J. D. and A. R. Egan (eds.). 1975. *Trace Elements in Soil-Plant-Animal Systems.* New York: Academic Press, Inc.
Pais, I. and J. B. Jones, Jr. 1996. *Trace Elements in the Environment.* Boca Raton, FL: St. Lucie Press.
Peck, T. R., J. T. Cope, Jr., and D. A. Whitney (eds.). 1977. *Soil Testing: Correlating and Interpreting the Analytical Results.* Special Publication 29. Madison, WI: American Society of Agronomy.
Rechcigl, J. E. (ed.). 1995. *Soil Amendments and Environmental Quality.* Boca Raton, FL: Lewis Publishers.
Reid, K. (Ed.). 1998. *Soil Fertility Handbook.* Toronto: Ontario Ministry of Agriculture, Food and Rural Affairs, Queen's Printer for Ontario.
Sillanpää, M. 1972. *Trace Elements in Soils and Agriculture.* Soils Bulletin 17. Rome: FAO of the United Nations.
Sillanpää, M. 1990. *Micronutrient Assessment at the Country Level: An International Study.* Soils Bulletin 63. Rome: FAO of the United Nations.
Sillanpää, M. and H. Jansson. 1992. *Status of Cadmium, Lead, Cobalt, and Selenium in Soils and Plants of 30 Countries.* Soils Bulletin 65. Rome: FAO of the United Nations.
Sprague, H. B. (ed.). *Hunger Signs in Crops.* Third Edition. New York: David McKay Company.
Tan, K. H. 1998. *Principles of Soil Chemistry.* Third Edition. New York: Marcel Dekker.

ANALYTICAL TEXTS

Anon. 1986. *The Analysis of Agricultural Materials.* Third Edition. Reference Book 427. London: Her Majesty's Stationery Office.
Boumans, W. J. M. (ed.). 1987. *Inductively Coupled Plasma Emission Spectroscopy, Part II: Applications and Fundamentals.* New York: John Wiley & Sons.
Cunniff, P. A. (ed.). 1999. *Official Methods of Analysis of the Association of Official Analytical Chemists.* 16th Edition. Arlington, VA: Association of Official Analytical Chemists.
Ewing, G. W. (ed.). 1990. *Analytical Instrumentation Handbook.* New York: Marcel Dekker.
Lajunen, L. H. J. 1992. *Spectrochemical Analysis by Atomic Absorption and Emission.* Cambridge, England: The Royal Society of Chemistry.

Metcalfe, E. 1987. *Atomic Absorption and Emission Spectroscopy.* New York: John Wiley & Sons.
Mitchell, R. L. 1964. *The Spectrochemical Analysis of Soils, Plants, and Related Materials.* Technical Communications 44A. Bucks, England: Commonwealth Agricultural Bureaux, Farnham Royal.
Montasar, A. and D. W. Golightly (eds). 1987. *Inductively Coupled Plasma in Analytical Atomic Spectrometry.* New York: VCH Publishers.
Rump, H. H. and H. Krist. 1988. *Laboratory Manual for the Examination of Water, Waste Water, and Soil.* New York: VCH Publishers.
Ruzicka, J. and E. H. Hansen. 1988. *Flow Injection Analysis.* Second Edition. New York: John Wiley & Sons.
Smith, K. A. (ed.). 1983. *Soil Analysis: Instrumental Techniques and Related Procedures.* New York: Marcel Dekker.
Smith, K. A. (ed.). 1991. *Soil Analysis: Modern Instrumental Techniques.* 2nd ed. New York: Marcel Dekker.
Smoley, C. K. 1992. *Methods for the Determination of Metals in Environmental Samples.* Boca Raton, FL: CRC Press.
Thompson, M. and J. N. Walsh. 1983. *A Handbook of Inductively Coupled Plasma Spectrometry.* Glasgow: Blackie & Son.
van Loon, J. C. 1985. *Selected Methods of Trace Metal Analysis; Biological and Environmental Samples.* New York: John Wiley & Sons.
Varma, Asha. 1991. *Handbook of Inductively Coupled Plasma Atomic Emission Spectroscopy.* Boca Raton, FL: CRC Press.
Walsh, L. M. (ed.). 1972. *Instrumental Methods for Analysis in Soils and Plant Tissue.* Revised edition. Madison, WI: Soil Science Society of America.

QUALITY ASSURANCE

Dux, J. P. 1986. *Handbook on Quality Assurance for the Analytical Laboratory.* New York: Van Nostrand Reinhold.
Garfield, F. M. (ed.). 1984. *Quality Assurance Principles for Analytical Laboratories.* Arlington, VA: Association of Official Analytical Chemists.
Garner, W. Y. and M. S. Barge (eds.). 1988. *Good Laboratory Practices.* ACS Symposium Series 369. Washington, DC: American Chemical Society.
Taylor, J. K. 1987. *Quality Assurance of Chemical Measurements.* Chelsea, MI: Lewis Publishers.

Index

A

Acids, 228
Adams-Evans lime buffer, 41–45
Ammonium, 149
   2 N potassium chloride extraction, 152–154
Analytical methods, 207–224
   atomic absorption spectrophotometry (AAS), 213–214
   flame emission spectrophotometry, 212–213
   inductively coupled plasma emission (ICP) spectrometry, 214–219
   ion chromatography (IC), 219–220
   specific ion electrodes, 220–221
   segmented and continuous flow analysis (CFA), 211–212
   UV-VIS spectrophotometry, 209
Atomic absorption spectrophotometry (AAS), 213–214
Automated UV-VIS spectrophotometry, 211–212

B

Bases, 228
Blanks, 233
Boron,
   hot water extraction, 118–120
   Mehlich No. 3 extraction, 123–127
   organic soils and soilless growth media, 185–190
Buffer solutions, 228–229

C

Calcium
   Mehlich No. 1 extraction, 97–100
   Mehlich No. 3 extraction, 100–104
   Morgan extraction, 104–107
   neutral normal ammonium acetate extraction, 93–97
   organic soils and soilless growth media, 185–190
   water extraction, 109–114

Chloride, 165–166
   saturated calcium hydroxide extraction, 170–172
   0.01 M calcium nitrate extraction, 166–168
   0.5 M potassium sulfate extraction, 168–170
Conductance, 57
Conductivity
   1:1 soil:water solution, 57–61
   1:2 soil:water solution, 61–63
   saturated paste, 64–66
Conversion tables, 236–240
Copper
   ammonium bicarbonate-DTPA extraction, 127–130
   DTPA extraction, 1301–33
   Mehlich No. 3 extraction, 123–127
   organic soils and soilless growth media, 185–190

E

Exchangeable Acidity
   $BaCl_2$-TEA Buffer, 35–36
   1 N potassium chloride, 36–37
   Mehlich method, 149–154
Exchangeable aluminum, 36–37
Extractable reagents, 22–23

F

Flame emission spectrophotometry, 212–213

H

Heavy metals, 139
   ammonium bicarbonate-DTPA extraction, 139–142
   DTPA extraction, 142–145

I

Indicators, 227–228

245

Inductively coupled plasma emission (ICP) spectrometry, 214–219
Internal standards, 234
Ion chromatography (IC), 219–220
Iron
    Ammonium bicarbonate-DTPA extraction, 127–130
    DTPA extraction, 130–133
    Mehlich No. 3 extraction, 123–127
    organic soils and soilless growth media, 185–190

L

Lime requirement, 41
    Adams-Evans method, 41–45
    Mehlich buffer-pH method, 49–54
    SMP lime buffer—original and double-buffer, 45–49
Long-term storage, 23

M

Magnesium
    Mehlich No. 1 extraction, 97–100
    Mehlich No. 3 extraction, 100–104
    Morgan extraction, 104–107
    neutral normal ammonium acetate extraction, 93–97
    organic soils and soilless growth media, 185–190
    water extraction, 109–114
Major cations, 93
Manganese
    ammonium bicarbonate-DTPA extraction, 127–130
    DTPA extraction, 130–133
    Mehlich No. 3 extraction, 123–127
    organic soils and soilless growth media, 185–190
Matrix effects, 232–233
Mehlich buffer-pH method, 49–54
Micronutrients, 117–118
    boron, 118–120, 185–190
    copper, 123–130, 185–190
    iron, 123–130, 185–190
    manganese, 123–130, 185–190
    zinc, 120–130, 185–190

N

Nitrate, 149
    ammonium bicarbonate-DTPA extraction, 149–151
    organic soils and soilless growth media, 185–190
    2 N potassium chloride extraction, 152–154
    0.01 M calcium sulfate and 0.04 M ammonium sulfate extraction, 154–156
North American Proficiency Testing Program (NAPT), 7–10

O

Organic and humic matter, 175
    humic matter by 0.2 N NaOH extraction, 180–182
    loss-on-ignition, 178–179
    wet digestion, 175–178
Organic soils and soilless growth mnedia, 185–190

P

Specific ion electrodes, 220–221
Phosphorus, 69–70
    ammonium bicarbonate-DTPA extraction, 85–88
    Bray P1 extraction, 70–73
    Mehlich No. 1 extraction, 76–79
    Mehlich No. 3 extraction, 80–83
    Morgan extraction, 83–85
    Olsen's sodium bicarbonate extraction, 73–76
    organic soils and soilless growth media, 185–190
Potassium
    ammonium bicarbonate-DTPA extraction, 107–109
    Mehlich No. 1 extraction, 97–100
    Mehlich No. 3 extraction, 100–104
    Morgan extraction, 104–107
    neutral normal ammonium acetate extraction, 93–97
    organic soils and soilless growth media, 185–190
    water extraction, 109–112

Q

Quality assurance, 193
    blind studies, 201–202
    control charts, 198–200
    documentation, 202–205
    quality assurance, 193–196
    statistical control, 196–198
    system audits, 202

Index

R

Reagents, 10, 23, 225–227
Reference methods, 5–7

S

Sample preparation, 18–20
Sampling, 17–18
   devices, 25
Scoops, 20–22
SMP lime buffer, 45–49
Sodicity, 57
Soil
   laboratory factors, 22–23
   laboratory sample preparation, 18–20
   long-term storage, 23
   sample aliquot determination, 20–23
   sampling, 17–18
   sampling devices, 25
Soil testing
   changing role and needs, 3–5
   history, 1–3
   reference methods, 5–7
Soil pH, 27–28
   0.01 M calcium chloride, 31–33
   1 N potassium chloride, 33–35
   water, 28–30
Soluble salts, 57
   specific conductance in supernatant 1:2 soil:water solution, 57–61
   specific conductance in supernatant 1:1 soil:water solution, 61–63
   saturated paste method, 64–66
Sodium
   Mehlich No. 1 extraction, 97–100
   Mehlich No. 3 extraction, 100–104
   neutral normal ammonium acetate extraction, 93–97
   water extraction, 109–112
Specific ion electrodes, 220–221
Standards, 10, 23, 227
   preparation, 231–236
Standard acids, 228
Standard bases, 228
Standard preparation, 231–235
Standard sources, 236
Sulfate-sulfur, 159
   monocalcium phosphate (500 ppm) extraction, 159–161
   0.5 M ammonium acetate/0.25 M acetic acid extraction, 161–163

U

UV-VIS spectrophotometry, 209
   automated, 211–212

W

Water
   extraction, 109–112
   quality, 10–11

Z

Zinc
   ammonium bicarbonate-DTPA extraction, 127–130
   DTPA extraction, 130–133
   Mehlich No. 1 extraction, 120–123
   Mehlich No. 3 extraction, 123–127
   organic soils and soilless growth media, 185–190

9 780849 302053